THE NEXT GENERATION OF SCIENTISTS IN AFRICA

Edited by C. Beaudry, J. Mouton & H. Prozesky

AFRICAN MINDS

Published in 2018 by African Minds
4 Eccleston Place, Somerset West 7130, Cape Town, South Africa
info@africanminds.org.za
www.africanminds.org.za

ISBN Paper 978-1-928331-93-3
ISBN eBook 978-1-928331-94-0

African Minds is a not-for-profit, open-access publisher. In line with our goal of developing and fostering access, openness and debate in the pursuit of growing and deepening the African knowledge base and an Africa-based creative commons, this publication forms part of our non-peer reviewed list, the broad mission of which is to support the dissemination of knowledge from and in Africa relevant to addressing the social challenges that face the African continent.

Orders

African Minds
4 Eccleston Place, Somerset West 7130, Cape Town, South Africa
info@africanminds.org.za
www.africanminds.org.za

For orders from outside South Africa:
African Books Collective
PO Box 721, Oxford OX1 9EN, UK
orders@africanbookscollective.com
www.africanbookscollective.co

Acknowledgements

'The research system in Africa: perceptions, career development, support and performance of early-career scientists: A multinational study' project was made possible by funding from the Robert Bosch Stiftung. This work was carried out with the aid of a grant from the International Development Research Centre, Ottawa, Canada. The views expressed herein do not necessarily represent those of IDRC or its Board of Governors.

Contents

Preface

This book is based on a four-year study of Young Scientists in Africa (YSA). The study, which was jointly funded by the IDRC (Canada) and the Robert Bosch Stiftung (Germany), commenced in April 2015 and was completed in October 2018. The main focus of the study was to investigate the factors that influence the research performance and career development of young scientists in Africa. The research design for the study involved a multi-design approach with constitutive elements of bibliometric analyses of research data, a web-based survey of African scientists and more than 250 qualitative interviews with selected young scientists in Africa.

Higher education can have broad, positive effects on economic and social development, for both the individual and society in general, as it contributes to producing a skilled workforce, and the research performed within the higher education system generates knowledge, stimulates international cooperation and increases competitiveness in the global knowledge economy. This has been repeatedly documented by universities, foundations and international organisations (Kimenyi 2011; World Bank 2000, 2013). They show that a country's higher education system influences its capacity to find innovative solutions to societal problems and needs. A strong higher education system should respond to a country's needs by producing well-trained experts and creating conditions that inspire originality and productivity. A wider access to higher education has been suggested to be a precondition for enabling countries to take control of their own political agendas and, thus, to strengthen the legitimacy of their democratic governments.

The central role of universities holds even truer for the African continent, where universities do not have plentiful private research laboratories that are provided resources by government institutions. Investing in strong higher education systems and promoting the development of adequate career development and training opportunities for early-career scientists in Africa is thus expected to result in benefits for African societies at large.

In sub-Saharan Africa the gross enrolment ratio at tertiary level is only 6% (World Development Indicators 2011). Although this is very low compared to other regions, in the past decades the demand for higher education in Africa has undergone a significant increase (Mohamedbhai 2011, 2014). African universities have not been adequately prepared to accommodate the large numbers of students. Many universities have also not been able to provide strong support for the advancement of young doctoral graduates who need to cope with the demands for increasing quantity and quality of research activities in the competitive global research arena, but also to address the issues and challenges that are important to the region. Governmental financial support has generally not matched this increasing demand for education and research intensity, and universities are underfunded, resulting in the deterioration of infrastructure, the underdevelopment of essential services, as well as a critical shortage of academic staff.

This poses serious challenges to the African continent and its ability for transformation and development in the future, in spite of development goals and increased political commitment in this regard. Understanding the impact of funding and support, as well as the perceptions and opinions, of young scientists on their research productivity and performance will help to inform decision-makers in academia and politics in a meaningful and effective way about the most critical changes needed. Targeting young doctoral graduates is likely to have a long-lasting effect. Gaining insight regarding the research environment (composed of universities, granting councils, academies, science and technology policy, and so on) in which young scientists evolve is intrinsically linked to their well-being, performance and socio-economic impact. Future capacity building in Africa depends on our comprehension of the research system and on the ways to improve it.

Why focus on young scientists? Given their pivotal position, we would argue that knowledge about the state of early-career scientists in higher education in Africa holds an important key to understanding current challenges and achieving future success in this sector. Young scientists are a powerful resource for change and sustainable development, as they are at the heart of innovation and knowledge creation. The initial conditions of their career will likely affect their entire career path. Their opportunities for education, training and creative development determine how prepared the continent can be to face today's challenges, as well as those that will arise in the future. In general, the availability of literature and comparable findings on the state of young scientists is sparse. One of the few international comparative projects is the Carnegie Study on the Academic Profession (surveyed in 1992 and published between 1995 and 1998) (Altbach & Whitelaw 1994; Altbach 1996; Arimoto & Ehara 1996; Boyer, Maassen & Van Vught 1996; Teichler 1998).

In addressing a wide variety of research questions that guided the YSA study, the book aims to assist its readers in better understanding the African research system in general, and more specifically the challenges young scientists face with regard to their careers and research performance.

The book consists of three sections in addition to an introduction and technical appendix. Part One provides the bigger context for the study through a focus on the state of African science today. Part Two is devoted to the core research question of the study: What are the main challenges that young scientists in Africa face as far as their academic and scientific careers are concerned? Part Three focuses on two dimensions of research performance: The research output and research collaborations and networks as reported by our respondents. The final section of the book summarises the main findings and places some policy recommendations on the table.

The individual authors of each of the chapters are listed at the beginning of each chapter. The study was a team effort. We specifically want to acknowledge the contributions of Charl Swart at the Centre for Research on Evaluation, Science and Technology (CREST) who managed the survey work, Rein Treptow who managed the interview work at CREST and Pauline Huet who did the same in Montreal. Statistical contributions have been made by Carl St-Pierre in Montreal. In addition to acknowledging the valuable contributions of our team, we also need to express our gratitude to all the technical and administrative support staff of our respective units. These include those who conducted the interviews (at CREST: Marina Joubert, Milandre van Lill, Isabel Basson and Melissa Coetzee, and at Polytechnique Montreal: Michel Samy Diatta, Birné Ndour, Mehdi Rhaiem and Lamia Tazi Saoud), the project management team (at CREST: Marthie van Niekerk, Rolene Langford, Nigel Jansen and Lenny Poole, and at Polytechnique Montreal: Laurence Solar-Pelletier) and Lynn Lorenzen for information support.

We also want to express our thanks to our two international funders for their support: the IDRC and Robert Bosch Foundation.

And last, but not least, our sincere appreciation to the more than 7 500 scientists and scholars from across the African continent who took time to complete our survey, and more than 250 of those who participated in the follow-up qualitative interviews.

Catherine Beaudry
Johann Mouton
Heidi Prozesky

October 2018

List of acronyms and abbreviations

ACE	African Centres of Excellence
AGRIC	agricultural sciences
CWTS	Centre for Science and Technology Studies, Leiden University
CREST	Centre for Research on Evaluation, Science and Technology, Stellenbosch University
DFG	Deutsche Forschungsgemeinschaft (Germany)
ENG	engineering and applied technologies
GDP	gross domestic product
GERD	gross domestic expenditure on R&D
GloSYS	Global State of Young Scientists Project
HEALTH	health sciences
HSRC	Human Sciences Research Council (South Africa)
HUM	humanities
IDRC	International Development Research Centre (Canada)
MNCS	field-normalised citation score
NGO	non-governmental organisation
NPO	non-profit organisation
NIH	National Institutes of Health (US)
NORAD	Norwegian Agency for Development Cooperation
NS	natural sciences
OECD	Organisation for Economic Co-operation and Development
R&D	research and development
RFS	relative field strength
Sida	Swedish International Development Cooperation Agency
SS	social sciences
STEM	science, technology, engineering and mathematics
UIS	UNESCO Institute for Statistics
UNESCO	United Nations Educational, Scientific and Cultural Organisation
USAID	United States Agency for International Development
WHO	World Health Organisation
YSA	Young Scientists in Africa Project

About the editors

Catherine Beaudry is a Rhodes Scholar and holds a PhD in economics from the University of Oxford. From her first degree in electrical engineering specialised in satellite technology, she has kept a strong interest on technology, science and innovation. She is a professor at the Mathematics and Industrial Engineering Department of École Polytechnique de Montréal where she also holds a Tier I Canada Research Chair on the Creation, development and commercialisation of innovation. In addition, she is an adjunct professor at Université du Québec à Montréal (UQAM), a member of the Centre for Interuniversity Research on Science and Technology (CIRST), a fellow at the Centre for Interuniversity Research and Analysis of Organisation (CIRANO), and a founding member of the Global Young Academy (GYA). Her current research focuses on open innovation in the aerospace industry, on the impact of university funded research in biotechnology and nanotechnology as well as on the commercialisation of nanotechnology. In addition, her main research interests are the analysis of innovation networks, collaboration, partnerships and alliances, industrial clusters, and how they influence innovative firm performance and survival. Her expertise is the economics of science, technology and innovation using applied econometrics for its analysis.

Johann Mouton holds a PhD in Philosophy from the Rand Afrikaans University (University of Johannesburg subsequent to the mergers). He is professor in, and director of, the Centre for Research on Evaluation, Science and Technology (CREST) at Stellenbosch University and the DST-NRF Centre of Excellence for Scientometrics and STI Policy (SciSTIP). He is on the editorial board of five international journals including the *Journal of Mixed Methods Research, Science and Public Policy, Science, Technology and Society and Minerva.* He has authored or co-authored numerous books and articles on research methodology and postgraduate supervision. He has received two prizes from the Academy for Science and Arts in South Africa including one for his contribution to the promotion of research methodology in South Africa. In 2012 he was elected to the Council of the Academy of Science of South Africa. His main research interests are the philosophy and methodology of the social sciences, higher education knowledge production, sociology of science, scientometrics and science policy studies.

Heidi Prozesky holds a PhD in sociology on gender differences in the publication productivity of South African scientists. She has pursued this research interest since 2003 through her involvement in NACI-commissioned projects and supervision of students on the topic of women in academic science, and with a particular focus on scientometric analysis of research output and the sociology of science. From 2005 to 2013 she was a core team member – and in 2014 a research associate – of the DST–NRF Centre of Excellence in Invasion Biology. During this time, she developed a further specialisation in environmental sociology. Her third research interest is in social research methodology which she has been teaching at both undergraduate and postgraduate level for more than 15 years. In this field, she focuses on research design and proposal writing; questionnaire construction; issues affecting the validity of data collected though quantitative, qualitative and mixed methods designs; methodological trends in the social sciences; and research ethics. In 2015 she joined SciSTIP as a research manager.

List of authors

Catherine Beaudry Professor at the Mathematics and Industrial Engineering Department and Tier I Canada Research Chair, École Polytechnique de Montréal, Canada

Jaco Blanckenberg Postdoctoral research fellow at the Centre for Research on Evaluation, Science and Technology (CREST) and the DST-NRF Centre of Excellence in Scientometics and STI Policy (SciSTIP), Stellenbosch University, South Africa

Rodrigo Costas Senior researcher at the Centre for Science and Technology Studies (CWTS), Leiden University (the Netherlands) and Extraordinary associate professor at the Centre for Research on Evaluation, Science and Technology (CREST), Stellenbosch University

Csaba Kozma Research intern at the Centre for Science and Technology Studies (CWTS), Leiden University, the Netherlands

Agnes Lutomiah Doctoral student at the Centre for Research on Evaluation, Science and Technology (CREST), Stellenbosch University, South Africa

Clara Calero-Medina Researcher at the Centre for Science and Technology Studies (CWTS), Leiden University, the Netherlands

Johann Mouton Director of the Centre for Research on Evaluation, Science and Technology (CREST) and the DST-NRF Centre of Excellence in Scientometics and STI Policy (SciSTIP), Stellenbosch University, South Africa

Heidi Prozesky Associate professor at the Centre for Research on Evaluation, Science and Technology (CREST) and Research manager at the DST-NRF Centre of Excellence in Scientometics and STI Policy (SciSTIP), Stellenbosch University, South Africa

Charl Swart Postdoctoral research fellow at the Centre for Research on Evaluation, Science and Technology (CREST), Stellenbosch University, South Africa

Rein Treptow Researcher at Centre for Research on Evaluation, Science and Technology (CREST), Stellenbosch University, South Africa

PART ONE

Context: The state of science in Africa

CHAPTER 1

African science: A diagnosis

Johann Mouton

African science: A legacy of neglect

The decline of university research in Africa in the late 1990s and early millennium

Various international forces associated with globalisation and internationalisation of trade in the 1980s and 1990s have had a devastating effect on the economies of many African countries: The decline in export volumes as well as the relative decline in the price of primary products in world trade in the 1980s and 1990s, combined with the mishandling of exchange rates and of external reserves, and the huge external debt overhang together created major resource gaps for the countries of Africa. This put serious pressure on their import capacity and the availability of resources for essential economic and social investment. The result was an increased dependence of the typical sub-Saharan Africa country on aid from the developed countries.

At the same time, international agencies, most notably the World Bank, decided to privilege expenditure on basic education at the expense of support for higher education. This policy position was based on two premises. The first was the belief that the returns to investments in primary and secondary education are higher than those to higher education. The second reason related to concerns with equity and access to basic education which would naturally lead to an emphasis on primary or education. The result was quite predictable, with many universities thrown into financial crisis, laboratories and libraries not receiving any maintenance, overcrowded lecture rooms and huge flight of the top academics from these institutions.

Research and scholarship would be one of the main losers during these years. Africa's share of world science, as measured in papers published in the citation indexes of the Web of Science, declined steadily over this period. Bibliometric studies conducted at the University of Leiden's Centre for Science and Technology Studies showed that sub-Saharan Africa's share of world scientific papers declined from 1% in 1987 to 0.7% in 1996 (Tijssen 2007). This diminishing share of African science overall did not reflect a decrease in absolute sense, but rather an increase in publication output less than the worldwide growth rate. Africa had lost 11% of its share in global science since its peak in 1987; sub-Saharan science had lost almost a third (31%). The countries in Northern Africa (Algeria, Egypt, Mauritania, Libya, Morocco and Tunisia) accounted for the modest growth of the African share of the worldwide output during the years 1998 to 2002.

Bibliometric analysis of research output is only one measure of the relative decline of research and scholarship at many African universities. Numerous case studies covering the period between 1990 and 2005 demonstrated quite convincingly that research at former well-resourced and supported institutions, such as Makerere University in Uganda, Ibadan in Nigeria and University of Dar es Salaam in Tanzania, had deteriorated; that research infrastructure and the general state of laboratories at many institutions had suffered from a lack of maintenance and timely replacement of old equipment. In addition, the generally poor quality of library resources has not improved significantly (many university libraries do not have automated management systems in place); and the demand for sufficient funding for ongoing research and scholarship continues, as does the need for proper research management and support at most of these institutions.

The cumulative effect of the funding policies of the last two decades of the previous millennium, the huge growth in student enrolments in higher education institutions, combined with continuing political instability in many African countries, created a state of affairs which Mouton (2008) described as the 'de-institutionalisation' of science.

The de-institutionalisation of research institutions in Africa

Science systems in developed and highly industrialised countries have a certain number of clear and evident features. Such systems are dense (well-populated) with highly articulated scientific institutions. 'Scientific institution' is defined as any formal organisation or entity dedicated to the pursuit of scientific knowledge production, dissemination and utilisation. This definition includes bodies that perform R&D, such as university centers, laboratories and institutes, as well as R&D performing entities outside the higher education sector. But it also includes scientific publishing houses, journals, conferences, workshops and seminars which are 'organisations' for the dissemination of scientific knowledge. And it includes bodies such as technology incubators, technology transfer offices, patenting offices, and so on, that promote the utilisation and commercialisation of scientific knowledge.

In a modern science system there are typically a multitude of these scientific institutions that perform clearly articulated functions and roles, and together constitute what could be termed the 'national mode of scientific production' (Gaillard et al. 2002; Waast & Krishna 2003). The 'national mode' means that science is conducted for the public good and that the direction of science is shaped and steered by a nation's most pressing socio-economic needs. It also implies that the state accepts that it has a major responsibility for financing research and development activities.

Unfortunately, few or none of the features of modern science system applied to many countries in sub-Saharan Africa in the last two decades of the previous century. Many of the scientific institutions in these countries were fragile and susceptible to the vagaries of political and military events. They were severely under-resourced, and suffered because of a lack of clarity and articulation of science governance issues (demonstrated by constant shifts in ministerial responsibility for science).

De-institutionalised science exhibits five characteristics:

1. weak scientific institutions: fragile research centers and institutes, non-sustainable scientific journals, ineffectual scientific societies and academies of science;
2. dependence on international funding for R&D;

3. individualism in research rather than institution building;
4. inadequate reproduction of the scientific and academic work-force (decline in the number of doctoral programmes and doctoral students); and
5. weak inscription of science in African societies.

We argue below that this state of affairs has started to change in recent years. There is increasing evidence of small but robust institutions (some universities and research centres) that have survived the ruptures of political changes and economic fluctuations where pockets of significant science are now found. In these isolated cases (Botswana is a good example), science is publicly supported by the government, and there is reasonable political stability and good governance of the science system. In many of these cases, there are also well-established links and collaborative networks with strong research establishments elsewhere in the world.

But what were the factors that produced the (de) institutionalisation of science in many African countries especially between 1980 and 2000? At least six major factors contributed to this state of affairs:

1. the continuing legacy of colonial science in many countries;
2. the destabilising influence of political events and civil wars;
3. the devastating impact of World Bank policies on higher education in Africa;
4. the role of international agencies in shaping African sciences;
5. the continuing low investment in science by African governments; and
6. the continuing effects of the brain drain.

Colonial science legacy

Many of the research institutes that were established during colonial rule still exist in African countries. It is now well documented that the role of different colonial powers in the formation of scientific institutions varied greatly across continents. This is both a function of the nature of the institutions that were established, as well as the 'model' of 'colonial' science pursued.

What is perhaps not so clear is how the continuing legacy of colonial scientific institutions in many African countries should be assessed. On the one hand, such institutions had the negative effect of creating a long-term dependency, by the African country, on the colonial power long after independence, which led to a neglect in establishing local institutions (cf. Gaillard's [2003] thesis in this regard in his study of the Tanzanian science system). On the other hand, some of the institutes (such as the Pasteur institutes in Francophone countries) remain sites of significant capacity and provide a stabilising continuity within the scientific landscape of these countries.

Political stability and civil wars

The destabilising influence of many regional and local political events had led to the closing of scientific institutions (universities) in many countries and effectively put science back many decades. Events such as the civil war in Rwanda/Burundi, the Mengistu regime in Ethiopia, Amin's dictatorship in Uganda, as well as the civil wars in Mozambique and Angola, are examples. These events had widespread negative impacts on institution building in these countries. In many cases it led to the suspension of overseas research funding (e.g. Sida/SAREC suspending its support to Ethiopia in the late 1990s), the closing of institutions because of lack of government funding and, perhaps most notably, the flight of top academics and scientists to other parts of the world. A good example of the devastating impact on a single institution is that of the University of Makerere in Uganda. Once a major site for internationally recognised research in the 1950s and 1960s, it suffered because of civil war and lack of government funding in the 1980s and beyond. This has forced the University in the 1990s to take in many more students than it could support (in order to raise fee-income), with the result that, by the beginning of this millennium, it has more than 30 000 students for a campus built for less than 15 000. It is only in recent years that student growth has been capped and a decline in student numbers has materialised.

Structural adjustment policies and economic decline

Various international forces associated with globalisation and internationalisation of trade in the 1980s and 1990s also had a devastating effect on the economies of many African countries: The decline in export volumes as well as the relative decline in the price of primary products in world trade in the 1980s and 1990s, combined with the mishandling of exchange rates and of external reserves, and the huge external debt overhang, together created major resource gaps for the countries of Africa. This put serious pressure on their import capacity and the availability of resources for essential economic and social investments. The results included increased dependence of the typical sub-Saharan Africa countries on aid from the developed countries.

As formulated by Sawyer (2004):

> The collapse of many national economies in Africa under these forces and the accompanying destabilisation of social structures threw all institutions, including those of higher education, into a prolonged crisis. A variety of structural adjustment programmes (SAPs) were introduced in the 1980s and 1990s to reverse the economic and social crises. The programmes were intended, first, to give freer rein to market forces by removing rigidities in the production, pricing, marketing and exchange rate regimes. They also sought to cut back the role of the state, downsizing it and reducing its reach. All this was to be combined with the rapid opening up of the economy to international competition. The results are yet new challenges to Africa's universities – the downgrading of university funding (in favour of basic education) and the pressure on them to adjust to the severe austerity regimen imposed by the various economic stabilisation policies, at the same time as they were pressured to increase enrolment and maintain quality levels, without commensurate increases in resources ... A further factor was the policy of privileging expenditure on basic education at the expense of higher education, a posture reflecting the policy positions of the World Bank and leading donor agencies, and the argument that the social rate of return on investments in basic education was higher than in higher education.

To summarise, at the same time as university enrolments increased exponentially in many African countries, both government support and external donor aid to higher education was dramatically reduced. The result was quite predictable, with many universities thrown into financial crisis, laboratories and libraries not receiving any maintenance, overcrowded lecture rooms and the flight of the top academics from these institutions. It was only towards the end of the 1990s that these trends were being reversed and government and international aid to universities in Africa being restored.

International research and funding agencies

The role of international agencies in shaping and steering science on the African continent cannot be underestimated. In this regard we include both the role of international development and aid organisations such as Sida/SAREC, NORAD, Carnegie, Ford, Rockefeller, USAID, IDRC and many others, as well as the presence of international research bodies such as the CGIAR institutes, WHO research institutes and so on. It is clear that countries which house the headquarters of these organisations or significant institutes thereof, benefit immensely from their presence. The significance of international institutes is manifold:

- they provide some continuity in research programmes in the countries where they are located;
- they are conduits for R&D funding through their international donors;
- they form networks of collaboration and expertise that cut across national boundaries;
- they provide employment to local scientists in countries where research employment is limited; and
- they usually have much better facilities and laboratories for conducting research than the local universities and research institutes of the host country.

On the downside, except in very general terms, one could not speak of a close alignment between the research priorities and programmes of these institutes and the national R&D priorities of individual countries. These institutes do not fall under the governance of the national science system of the host country and cannot be said to contribute in any strong sense to national institution-building. The research agendas and priorities of these institutes are usually set at a supra- or international level. So, although their presence has had a positive impact on science in those countries and in the regions (and there have been well-documented success stories), in the final analysis they remain disconnected from the 'national science systems' of these countries.

The continuing effects of the brain drain

Arguably, the biggest cause of the decline of African science during the 1980s and 1990s was the devastating effects of the erosion of human capital through the brain drain. Studies sponsored by the Research and Development Forum for Science-Led Development in Africa (RANDFORUM) reveal that up to 30% of African scientists – that is, excluding other professionals – were lost due to the brain drain. According to the Economic Commission for Africa (UNECA) and the International Organisation for Migration (IOM), an estimated number of 27 000 skilled Africans left the continent for industrialised countries between 1960 and 1975. Since 1990, at least 20 000 qualified people have left Africa every year. This means Africa has 20 000 fewer people who can deliver public services and articulate calls for greater democracy and development (Nunn 2005).

Ultimately, the restoration and improvement of African research institutions, and specifically African universities, would require strategies that focus on institution-building interventions rather than on building the capacity of individual scientists. This does not mean that training of, and support to, individual scientists, whether they are emerging or established scientists, is unimportant. On the contrary, we would argue that individual capacity-building is essential, but that it should be embedded in a framework of building the institutions of science in Africa.

A new narrative: Africa rising?

Around the turn of the millennium a new narrative began to emerge. Not surprisingly, this was reflected in the media headlines. Both in political culture and economic growth, it became clear that many African countries had started to turn the tide. An increasing number of African countries embarked on establishing democratic systems of governance. Africa's economic performance in recent years has given rise to the hope that political maturity and stability are being translated into economic growth.

In a study conducted for the Higher Education division of the World Bank in 2012, CREST identified nearly 120 initiatives to strengthen higher education in Africa over the past 30 years (with significant increases in these programmes since 2000). The study distinguished between three phases:

1. 1st generation programmes (1970 to 1990): Programmes in the fields of agriculture, and to some extent health sciences, were the primary determinants for international funding.
2. 2nd generation programmes: From the 1990s onwards the interest shifted to programmes aimed at the 'new' diseases (HIV/AIDS, Malaria/TB) as well as new disciplines such as economics.
3. 3rd generation programmes: Since the turn of the century we have witnessed a new focus on 'neglected' disciplines such as mathematics, physics and also the social sciences and humanities, as well as emerging applied innovation-related disciplines (biotechnology and materials science).

New investments in African science

Various international agencies have, in recent years, committed to increasing their investments in African science. An example is the World Bank that announced on 15 April 2014 that they have approved US$150 million to finance 19 university-based centers of excellence in seven countries in West and Central Africa. These competitively selected centers will receive funding for advanced specialised studies in science, technology, engineering and mathematics (STEM)-related disciplines, as well as in agriculture and health sciences. This landmark Africa Centres of Excellence (ACE) project is being financed through International Development Association (IDA) credits to the governments of Nigeria

(US$70 million), Ghana (US$24 million), Senegal (US$16 million), Benin, Burkina Faso, Cameroon, and Togo (US$8 million each). The Gambia will also receive a US$2 million credit and a US$1 million grant to provide higher education, including short-term training, to students, faculty and civil servants through the 19 ACEs.

Another interesting trend has been recent investments in 'big science'. What is significant about many of these new initiatives is the emphasis on strengthening African science institutions.

- Investments by agencies such as IDRC and DFID in national organisations such as science granting councils to increase the capability of states to plan science better, disburse research funds fairly and equitably and monitoring the value and impact of scientific research in African countries.
- Investments by the World Bank, the Wellcome Trust, and others, in research chairs and centres of excellence to create sustainable research performing units.
- Investments in doctoral programmes and training to build and expand the future generation of African scientists and academics (DAAD, Carnegie, Ford, Wellcome Trust, Gates Foundation and many others).
- Investments in scientific databases and scientific journals to increase the visibility of African science (Carnegie).

Investment in R&D and continued reliance on foreign funding

Many African governments have committed themselves to increasing their gross domestic expenditure on R&D (GERD). GERD is generally regarded as a measure of how dedicated a specific country is to supporting research. But the reality is that most sub-Saharan Africa countries spend less than 0.5% of their gross domestic product (GDP) on R&D. Nigeria, for example, lags far behind, in that only 0.2% of its GDP is assigned towards the development of R&D (African Innovation Outlook 2010: 37). Unfortunately, not all sub-Saharan Africa countries' GERD is captured in the statistics below (Table 1). We lack therefore a comprehensive view of GERD in the region. In South Africa, the R&D Survey notes that 'this ratio (GERD/GDP) has stagnated between 1.4 and 1.5 over the previous seven years' (HSRC 2014: xiii).

Table 1: **Gross domestic expenditure on R&D (GERD)**

Country	African Innovation Outlook				UNESCO° Institute for Statistics
	Year	GERD Million PPPS	GERD per capita PPPS	GERD as % of GDP	GERD as % of GDP
Botswana	2005	n/a	n/a	0.38	0.52 (2005)
Burkina Faso	2009	n/a	n/a	0.18	0.20 (2009)
Cameroon	n/a	n/a	n/a	n/a	n/a
Côte d'Ivoire	n/a	n/a	n/a	n/a	n/a
Ethiopia	2005	n/a	n/a	0.2	0.24 (2010)
Ghana	2008	78.7	58.3	0.47	0.23 (2007)
Kenya	2007	277.8	7.4	0.38	0.42 (2007)
Malawi	2007	180.1	12.9	1.70	n/a

Country	African Innovation Outlook				UNESCO◊ Institute for Statistics
	Year	GERD Million PPPS	GERD per capita PPPS	GERD as % of GDP	GERD as % of GDP
Mozambique*‡	2007	42.9	2.0	0.25	0.47 (2010)
Namibia	2005	n/a	n/a	0.3	n/a
Nigeria*†	2007	583.2	3.9	0.20	0.22 (2007)
Senegal	2008	99.0	8.0	0.48	0.37 (2008)
South AfricaΩ	2010/11	4 976.6	102.4	0.76	0.87 (2009/10)
Tanzania*	2007	234.6	5.8	0.48	n/a
Uganda†	2007	359.8	11.6	1.10	0.41 (2009)
Zambia	2008	55.3	4.6	0.37	0.34 (2008)
Zimbabwe	2005	n/a	n/a	0.2	n/a

Source: African Innovation Outlook (2010: 34)

* Data do not include the business enterprise sector
† Data do not include private non-profit institutions/organisations
‡ Data do not include the higher education sector
Ω HSRC CESTII Report (August 2013)
◊ We have added an additional column to include the latest available UNESCO Institute for Statistics statistics on R&D investment for selected countries

Figure 1: **Global R&D Expenditure by region (2011) – billions of US PPP dollars**

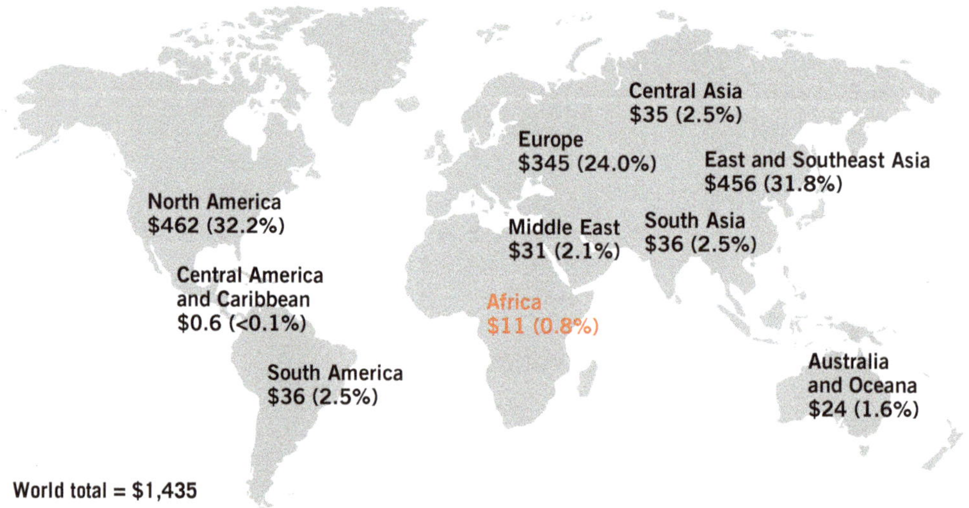

PPP = purchasing power parity.
Notes: Foreign currencies are converted to US dollars through PPPs. Some country figures are estimated. Countries are grouped according to the regions described by *The World Factbook*, available at www.cia.gov/library/publications/the-world-factbook/index.html.

It is clearly not the case that African governments are investing more in R&D and science than in the previous millennium. The latest figures from the Organisation for Economic Co-operation and Development (OECD) shows that government expenditure on R&D (GERD) as proportion of GDP remains the lowest of all regions in the world. Only a few countries in Maghreb and South Africa spend around 0.7–0.8% of GDP on R&D. For the majority of the countries, this proportion is below 0.4%. This is despite an explicit commitment by the ministers of science and technology in 2005 to aim for 1.00%.

The second Africa Innovation Outlook report (2014) reported Africa's GERD by funding sources for six countries that completed the R&D survey for the report. Government funding of R&D activities is significant, albeit at very low levels in real money terms. Notably, Ghana's government expenditure in R&D is the highest, accounting for 68% of its research expenditure. Ghana also records the lowest expenditure by its business sector, at 0.1% in 2010. In the majority of the six countries, contributions from the business/private sector are low. The outlier is South Africa, where the private sector contributes more than 40% of the total R&D expenditure. South Africa was also the least reliant on foreign funding, with only 12% of funding being from outside sources. The Africa Outlook report indicates that some countries, such as Mozambique, Burkina Faso and Uganda, received more than 50% of their R&D funding from foreign sources (Figure 2).

Figure 2: **Proportion of international funding for R&D by country (2010 or latest year)**

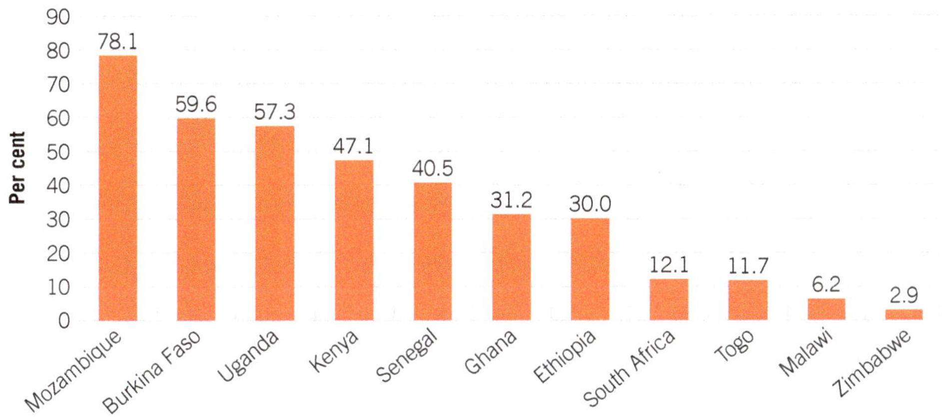

Source: ASTII R&D surverys 2010 or latest year available

To illustrate this over-reliance on foreign funding, Figure 3 reflects the sources of funding for Makerere University, Uganda's flagship university (Hydén 2017: 97).

Of all of Makerere's funders over the 12-year period between 2000 and 2012, the Uganda National Council for Science and Technology (UNCST) was the only local funder.[1] Makerere University has sustained much of its research activities through the assistance of external funders, among which are two European countries (Norway and Sweden), the USA, two foundations and the EU.

A cautionary note: We need to be critical and even sceptical of the validity of the new narrative. Even if it is the case that there has been a shift in political culture amongst some African nations, the spread of democratic forms of governance is not universal. There are still many cases (Zambia and Uganda) where lifelong presidents preside. There are still

1 Over this 12-year period, UNCST contributed a total of US$1 245 898 directly and also US$ 2 134 453 through the Millennium Science Initiative (MSI), a programme funded mainly by/through the World Bank (2016).

Figure 3: **Source of R&D funding at Makerere University (2000–2012)**

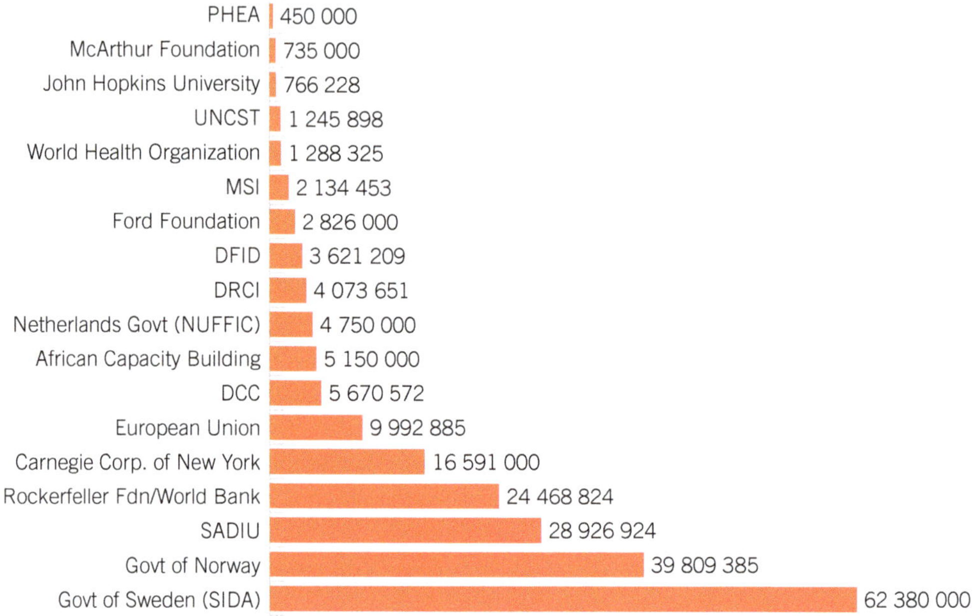

Note: PHEA = Partnerships for Higher Education in Africa; UNCST = Uganda National Council for Science and Technology; MSI = Millenium Science Initiative of the UNCST; DFID = Department for International Development, UK; IDRC = International Development Research Centre; CDC = Centers for Disease Control and Prevention; USAID = US Agency for International Development
Data Source: Directorate of Quality Assurance, Makerere University

pockets of conflict and ethnic 'war' in areas such as Southern Sudan, DRC and the terror activities of Bokom Aram continue to impact negatively on the lives of many peoples across the country. As far as turning the economic tide, this is again not a universal phenomenon. It is not the case that all African economies are growing (South Africa is effectively in a recession). More importantly, where economic growth is occurring it is mostly due to growth in extractive industries and the gains from increases in exporting commodities such as copper, zinc and other precious metals.

In the next chapter we turn our attention to some of the positive indicators of the rising tide of African science, including increased publication output, more international collaboration and increased mobility of African scientists. But it is not the case that these more positive trends necessarily reflect the impact of deliberate interventions and strategies of many African states. In fact, one could – quite cynically – claim that many of these more positive developments are occurring outside (and even despite) the decisions and funding of science and innovation by many African governments. The positive trends in scientific publication output are more likely due to (a) the impact of specific incentive schemes and directed funding of specific initiatives; (b) an increase in the number of African journal titles in the Web of Science (thus resulting in better coverage of African science); (c) the continued increase in investment by international funders; and (d) the accumulative effect of increased international collaboration between foreign scientists and African scientists in multi-authored teams in such fields as high-energy physics, infectious diseases and tropical medicine.

CHAPTER 2

African science: A bibliometric analysis

Johann Mouton and Jaco Blanckenberg

Introduction

In the previous chapter we presented a high-level overview of African science. We referred to the neglect of African science and higher education during the last three decades of the previous millennium. We also discussed the more recent revival and 'rising' of African science. These developments are evident in trends such as increasing international funding for certain areas of scientific research and university research in many African countries, the increasing investment in centres of excellence in certain fields, more funding for international collaborative initiatives, bilateral mobility agreements, and research capacity building. Understanding the larger landscape of African research and current trends is important when we want to understand the viewpoints, experiences and concerns of the next generation of African scientists. In this chapter we present a more granular picture of the research landscape in Africa. We employ standard bibliometric indicators to discuss:

- research publication output;
- the relative field strength of scientific fields;
- trends in research collaboration; and
- trends in citation impact.

Publication output

Tijssen (2007) showed how Africa's annual research output – as measured by articles in the Web of Science – was stagnating at that time. More recent studies have shown that the tide has since turned and that the number of African-authored papers have started to increase (AOSTI 2010; Mouton & Boshoff 2010). Updating these reports with our analysis of African-authored papers in the Web of Science shows how annual output has been increasing steadily over the past decade: from 15 285 in 2005 to 54 069 in 2016 (Figure 4). What is perhaps most striking is that this rate of increase has surpassed the world growth rate over the same period, with the result that Africa' share of world publication output more than doubled from 1.5% in 2005 to 3.2% in 2016.

Figure 4: **Africa world share and number of scientific publications (articles and reviews only): 2005–2016**

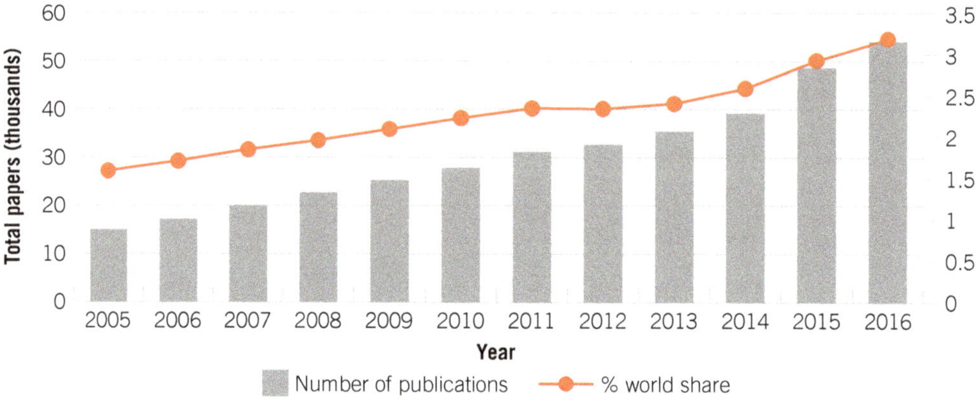

Publication output by country

Annual article output by country shows the continued dominance of South Africa, followed by strong contributions from Egypt and other Maghreb countries (Tunisia, Algeria and Morocco), together with smaller but significant contributions from Nigeria and the three Eastern African countries (Kenya, Uganda and Tanzania). The data also show how skewed the distribution of publication production on the African continent is. Thirteen countries that have each contributed 1% or more of total output in the most recent five-year period, account for 89% of all output. Table 2 lists the number of publications per country for two periods (2005 to 2010 and 2011 to 2015). We also present information on each country's share of Africa's total publication output for each time period. The final column indicates by colour whether the country's share improved (green) from the early to later period; whether it declined (red); or whether it stayed the same (yellow).

Table 2: **Country shares of Africa's publication production**

Country	2005–2010		2011–2015		Change in share
	No. of pubs	Country share	No. of pubs	Country share	
South Africa	53 072	29.1%	77 687	28.2%	
Egypt	32 267	17.7%	54 000	19.6%	
Tunisia	16 546	9.1%	25 420	9.2%	
Algeria	10 519	5.8%	18 313	6.6%	
Nigeria	13 583	7.5%	16 717	6.1%	
Morocco	9 295	5.1%	14 140	5.1%	
Kenya	6 954	3.8%	9 767	3.5%	
Uganda	3 666	2.0%	5 651	2.1%	
Ethiopia	2 934	1.6%	5 569	2.0%	
United Republic of Tanzania	3 707	2.0%	5 034	1.8%	
Ghana	2 832	1.6%	4 962	1.8%	

Country	2005–2010		2011–2015		Change in share
	No. of pubs	Country share	No. of pubs	Country share	
Cameroon	3 441	1.9%	4 463	1.6%	
Senegal	1 877	1.0%	2 635	1.0%	
Sudan	1 438	0.8%	2 393	0.9%	
Malawi	1 549	0.9%	2 356	0.9%	
Zimbabwe	1 691	0.9%	2 137	0.8%	
Burkina Faso	1 379	0.8%	1 938	0.7%	
Zambia	1 190	0.7%	1 853	0.7%	
Benin	1 051	0.6%	1 650	0.6%	
Botswana	1 370	0.8%	1 604	0.6%	
Libya	1 046	0.6%	1 496	0.5%	
Côte d'Ivoire	1 169	0.6%	1 471	0.5%	
Madagascar	1 021	0.6%	1 333	0.5%	
Mozambique	689	0.4%	1 198	0.4%	
Reunion	790	0.4%	1 108	0.4%	
Mali	1 009	0.6%	1 077	0.4%	
Rwanda	407	0.2%	1 068	0.4%	
Namibia	552	0.3%	931	0.3%	
Mauritius	460	0.3%	817	0.3%	
Democratic Republic of the Congo	391	0.2%	769	0.3%	
Gabon	624	0.3%	738	0.3%	
Gambia	686	0.4%	730	0.3%	
Congo	610	0.3%	715	0.3%	
Niger	468	0.3%	664	0.2%	
Togo	307	0.2%	435	0.2%	
Sierra Leone	106	0.1%	360	0.1%	
Guinea	182	0.1%	326	0.1%	
Angola	169	0.1%	314	0.1%	
Swaziland	195	0.1%	261	0.1%	
Seychelles	153	0.1%	212	0.1%	
Burundi	87	0.0%	169	0.1%	
Lesotho	135	0.1%	167	0.1%	
Mauritania	136	0.1%	163	0.1%	
Liberia	39	0.0%	146	0.1%	
Chad	113	0.1%	136	0.0%	
Eritrea	164	0.1%	114	0.0%	
Cape Verde	47	0.0%	109	0.0%	
Djibouti	23	0.0%	59	0.0%	
Somalia	11	0.0%	57	0.0%	
Comoros	22	0.0%	28	0.0%	
South Sudan	2	0.0%	7	0.0%	
Western Sahara	3	0.0%	1	0.0%	
	182 177		**275 468**		

In order to correct for the size of the countries, we divided the number of publications by the size of the population for each relevant year. We again compare the countries on this normalised indicator, for the same two periods (Figures 5a and 5b). The comparison between the two time periods, reveals some shifts in the rank of countries according to per capita number of publications (Figure 6).

Figure 5a: **Normalised output by country (2005–2010)**

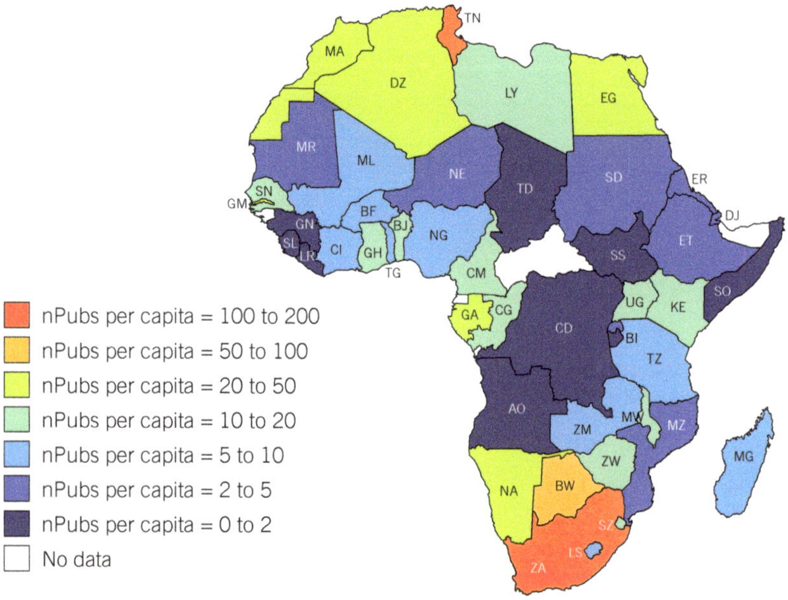

Figure 5b: **Normalised output by country (2011–2015)**

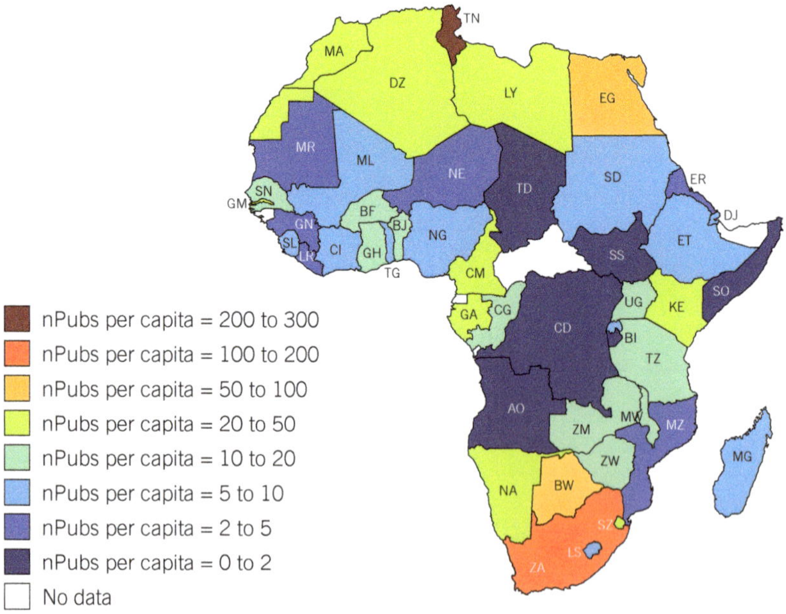

Figure 6: **Output per capita publications (2005–2010 and 2010–2015)**

Output by scientific field

Africa's production is higher than the overall average (3.2%) in 86 of the total of 273 subject categories in the Web of Science. If one focuses on those fields that are both large in volume (more than 5 000 papers produced between 2005 and 2015) and contribute significantly to world output (all more than 4%), nine fields meet these criteria: tropical medicine; parasitology; infectious diseases; public, environmental and occupational health; water resources; ecology; immunology; zoology; and plant sciences. These results reaffirm the fact that scientific production often mirrors the material reality of a country or region. In this case, the biodiversity on the continent, as well as the imperative to invest much effort in studying the (tropical) and other diseases that plague many African countries.

Table 3: **Scientific fields with highest contribution from Africa (in descending order for world share in 2015)**

Field	% World Share	Total Npubs (2005–2015)
Tropical medicine	24.71	7 380
Parasitology	15.85	6 643
Infectious diseases	13.22	13 183
Biomedical social sciences	8.01	1 889
Entomology	7.29	3 596
Ornithology	7.19	721
Andrology	6.77	250
Integrative & complementary medicine	6.41	1 770
Public, environmental & occupational health	6.28	13 729
Agronomy	6.11	4 697
Virology	6.07	3 661
Planning & development	6.02	1 670
Soil science	5.62	1 827
Biodiversity conservation	5.60	2 139
Biodiversity & conservation	5.60	2 139
Mining & mineral processing	5.60	1 390
Agriculture, dairy & animal science	5.58	3 152
Mycology	5.54	831
Agricultural economics & policy	5.45	499
Medical ethics	5.38	291
Water resources	5.23	5 893
Crystallography	5.07	2 772
Ecology	4.98	7 689
Immunology	4.92	8 605
Zoology	4.90	5 443
Plant sciences	4.89	9 345
Agriculture	4.84	13 624
Chemistry, medicinal	4.80	5 612
Mineralogy	4.78	1 249
Area studies	4.61	1 622
Microbiology	4.59	7 281
Veterinary sciences	4.53	5 683
Health policy & services	4.45	2 171
Horticulture	4.43	1 431
Geosciences, multidisciplinary	4.37	6 807
Geology	4.36	7 713
Public administration	4.32	1 780
Anatomy & morphology	4.24	678
Marine & freshwater biology	4.03	4 010
Environmental sciences & ecology	4.01	18 421
Demography	4.00	401
Agricultural engineering	3.92	909

Field	% World Share	Total Npubs (2005–2015)
Environmental sciences	3.91	11 515
Thermodynamics	3.84	3 222
Obstetrics & gynecology	3.78	3 470
Food science & technology	3.78	7 611
Agriculture, multidisciplinary	3.72	2 999
Health care sciences & services	3.70	3 209
Chemistry, applied	3.70	4 248
Physics, particles & fields	3.68	2 563
Cultural studies	3.62	401
Forestry	3.58	1 683
Evolutionary biology	3.57	2 066
Pharmacology & pharmacy	3.55	12 348
Physics, nuclear	3.54	1 996
Anthropology	3.52	1 566
Paleontology	3.49	947
Astronomy & astrophysics	3.49	4 205
Materials science, textiles	3.42	611
Spectroscopy	3.39	1 933
Environmental studies	3.36	1 816
Toxicology	3.34	2 698
Respiratory system	3.33	2 488
Fisheries	3.33	1 261
Multidisciplinary sciences	3.22	9 160
Archaeology	3.18	912
Materials science, composites	3.18	806
Pediatrics	3.10	3 953
Industrial relations & labor	3.08	269
Allergy	3.05	424
Nuclear science & technology	3.03	2 404
Engineering, chemical	3.02	6 916
Remote sensing	2.99	717
Electrochemistry	2.97	2 840
Nutrition & dietetics	2.95	2 800
Medical laboratory technology	2.93	1 174
Life sciences & biomedicine – other topics	2.92	3 333
Biology	2.92	3 333
Physics, condensed matter	2.90	6 029
Social issues	2.88	447
Imaging science & photographic technology	2.88	570
Mathematics, applied	2.86	7 523
Chemistry, inorganic & nuclear	2.84	3 486
Energy & fuels	2.84	4 809
Literature, african, australian, canadian	2.82	116
Medicine, general & internal	2.80	6 242

Relative field strength analysis

One of the standard indicators used in bibliometric studies to measure whether a country (or region or institution) is particularly strong in a particular field, is the specialisation or activity index.[2] Because this index measures the 'relative' strength of a particular field or discipline compared to others, we refer to it as the 'relative field strength index'. Africa's relative field strength (RFS[3]) is in the natural and agricultural sciences – the only broad domains where the RFS Index value is above one. Africa is weakest in the broad domain of the humanities.

Figure 7: **Relative Field Strength Index of Africa**

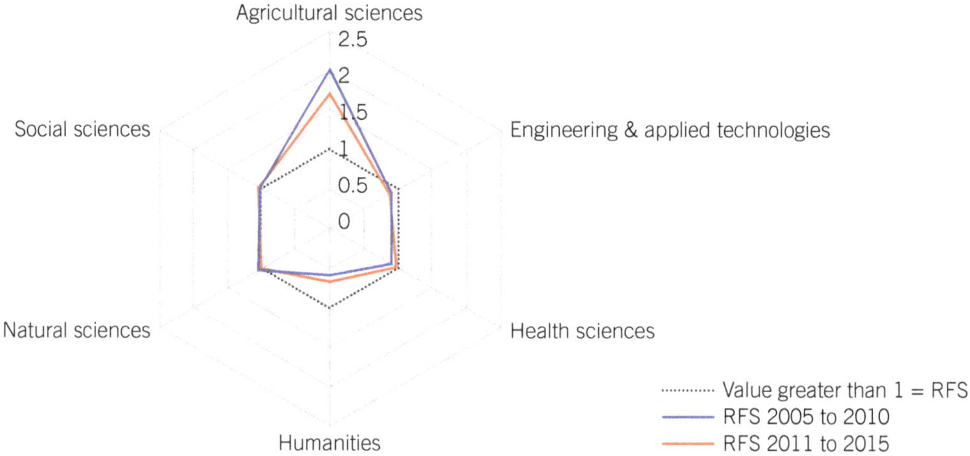

The disaggregation of the agricultural sciences by field shows that Africa is relatively strong in agricultural economics and policy (where it has increased its standing in the most recent period). Africa is also relatively strong in agronomy, plant sciences, and food science and technology – although its position in all three has weakened over time. In the broad area of the basic health sciences, it is clear that Africa is particularly active and strong in parasitology (a strength that has been sustained over the past ten years). Its relative activity in andrology and virology has also improved.

The disaggregation of the broad field of biological sciences (Figure 8) shows that Africa is particularly active and strong in entomology and ornithology and, to a lesser extent, zoology, biotechnology, and applied sciences and microbiology. However, in all of these fields, Africa has lost some ground in the more recent years. The disaggregation of the broad field of the physical sciences shows that Africa is relatively strong in three related fields (astronomy and astrophysics; particle physics; and nuclear physics). The disaggregation of the broad field

2 We are not convinced that the term 'activity' captures the notion of strength adequately. The term 'specialisation' is equally problematic, as it is more often used in discussions about specialisation within disciplines. Because this index measures the 'relative' strength of a particular field or discipline compared to others, we refer to it as the 'relative field strength index'.

3 A RFS value of 1 (indicated by the dotted line in all the graphs) in a field or discipline implies that this entity (country or region) has a world share for that field similar to its share in all fields combined. This is a 'neutral' situation, meaning there is no relative strength in that particular field. When the RFS value is greater than 1, the country is said to be strong in that field, at the expense of some other fields or disciplines for which the value is less than 1.

of the earth sciences shows that Africa is relatively strong across all subfields. Having said this, Africa has lost some ground in the most recent period in the fields of water resources, ecology and mineralogy. It remains strong in geology.

Figure 8: **Relative Field Strength Index: Biological sciences**

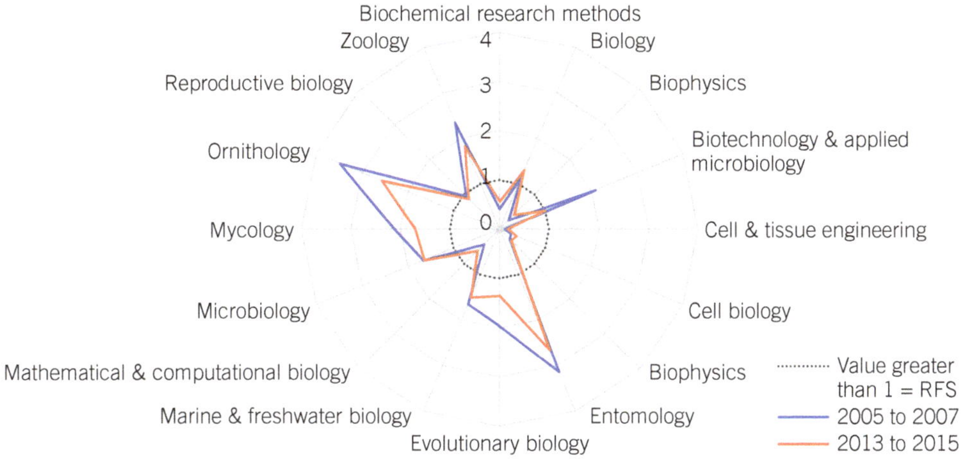

As far as the social sciences and humanities are concerned, the picture is less positive. The disaggregation of the broad field of psychology reveals that this is a general field in which Africa does not fare well. Like in many of the cases of the social sciences, this could be attributed to the neglect of the social sciences in most African countries (except for South Africa) as well as the under-representation of African social sciences/psychology journals in the Web of Science. The picture for sociology and related fields is very similar. Anthropology remains a relatively strong field whilst demography (strong in the earlier period), lost ground in the most recent period. The data for the disciplines in the field of language and linguistics paints a dismal picture. The only two fields that recorded some strength (but below the benchmark of 1) are linguistics and African literature. The results for other disciplines in the broad field of the humanities and arts shows small pockets of relative strength in medical ethics, cultural studies as well as archaeology.

Research collaboration

African scientists increasingly collaborate with scientists elsewhere.[4] Our analyses show that collaboration between countries on the continent only, is negligible. In addition, fewer than 10% of papers are single-institution (no collaboration) papers. The vast majority of Africa's papers fall into two categories: (1) papers of which the authors co-publish with institutions in the same country (national collaboration), which constitutes approximately 40% of all papers; and (2) papers involving some collaboration between Africa and the rest of the

4 It is standard practice in bibliometric studies to use the co-authorship relationships of journal articles as a measure of research collaboration. We analysed different categories of research collaboration (as measured by co-authorship patterns) in articles where there is at least one author from an African country.

world (international collaboration, which comprises approximately 50% of all papers). The trend is clearly in favour of the latter.

Figure 9: **Africa publication collaboration profiles (2005–2015)**

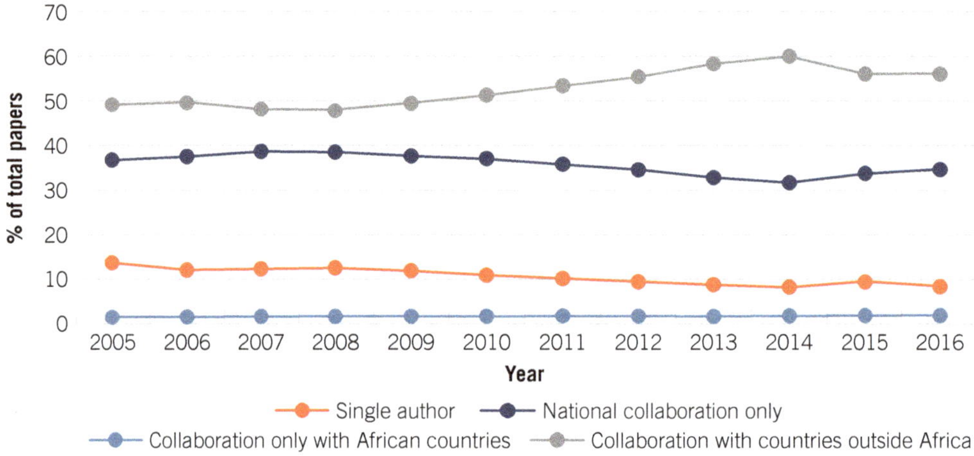

The increasing number of co-authorships (worldwide) should not be left without questioning. First we note large variations from one country to another. Not only do smaller countries tend to have higher collaborations, but collaborations are also related to policy choices. Thus, for instance, Turkey, China and Brazil have much lower levels of foreign co-authorship (Gaillard 2010), and in the three cases it is a choice to promote national publications that is reflected in the output measured by bibliometrics. Waast and Gaillard (2017) note that, in the last 30 years, shares of internationally co-authored papers have increased in very high proportions in all African countries. In some cases, as mentioned by Boshoff (2009), the proportion of foreign co-authored papers is very high (more than 80% in scientifically 'small' countries of Central Africa), and a survey of African co-authors showed they were rather in charge of empirical fieldwork and data collection (Boshoff 2009).

Another important element that needs to be emphasised is the emergence of large international endeavours translating into a high volume of publications with an extensive list of co-authors from various countries, most of whom do not even know each other or have never collaborated together. A large part of the increase of the production (yet to be calculated) relates to these 'big science' projects, such as international health projects in global health, or very large particle-physics projects such as ATLAS (Yami et al. 2011). Kahn (2018) shows that in the case of South Africa, the surge of international co-authorship is mostly the result of these collaborative projects rather than active cooperation and partnerships.

These mega projects blur the distinction between genuine collaborations among researchers and research units that indeed have worked together, and participation in large collaborative networks, funded globally, where units contribute to the knowledge base without necessarily being interconnected. Finally, from the point of view of researchers, in some cases international cooperation can be seen as 'time- consuming, costly and often one-sided', as was found to be the case in a survey of social sciences researchers in South Africa

(CREST 2014: 53). Although this might appear as strange and counterintuitive, it relates to the framework in which collaborations take place. Collaborations are welcome when they are the product of initiatives 'from the ground', or following tracks of well-established former contacts, and require no important additional effort. When levels of training and types of interrogations are very different, entailing some important effort in mutual understanding, collaborations will usually be rejected.

Trends in citation impact

The visibility of, and appreciation worldwide for, African science is partially captured by considering the number of times research publications are referenced ('cited') in the publications of other researchers working in the same discipline or related fields. The number of citations is partially dependent on the research fields (some fields are 'fast', others are 'slow'), and need to be corrected to allow fair cross-field comparisons. This is what the 'field-normalised citation score' (MNCS) does. Our analyses show that the citation impact of African-authored papers has been increasing steadily over the past 30 years – from 0.48 in 1980 to 0.73 in 2014. Although the overall impact is still below the gold standard of 1.0 (which would mean that it is generating similar citation scores than other countries or regions), it is the steady upward trend that is noteworthy.

Figure 10: **Trends in citation impact (MNCS) (1980–2014)**

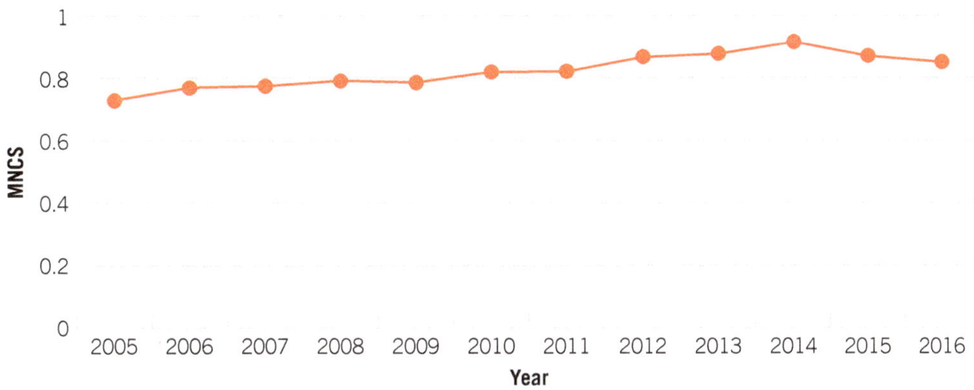

In order to identify high-impact fields, we selected only fields in which more than 1 500 articles have been published over this period. Given this threshold criterion, 15 fields recorded a MNCS of higher than 1. These are (in descending order from high to low impact): astronomy & astrophysics; nuclear physics; cardiac & cardiovascular systems; particles & fields physics; general & internal medicine; cardiovascular system & cardiology; energy & fuels; thermodynamics; electrochemistry; respiratory system; mechanics; biodiversity & conservation; genetics & heredity; evolutionary biology; and microbiology.

Our final bibliometric analysis – a positional analysis – combines two variables: the citation impact of a field or subfield with the score on the relative field strength index. We are essentially identifying those fields that are both strong (in relative world share) and highly visible. The two-dimensional diagram below plots fields on these two axes. The fields that score high on both of these two indicators would typically be in the top right-hand quadrant. The size of the bubble is commensurate to the number of publications in that field – the larger the bubble the bigger the volume of output.

Figure 11: **Africa 2010 citation impact vs relative field strength**

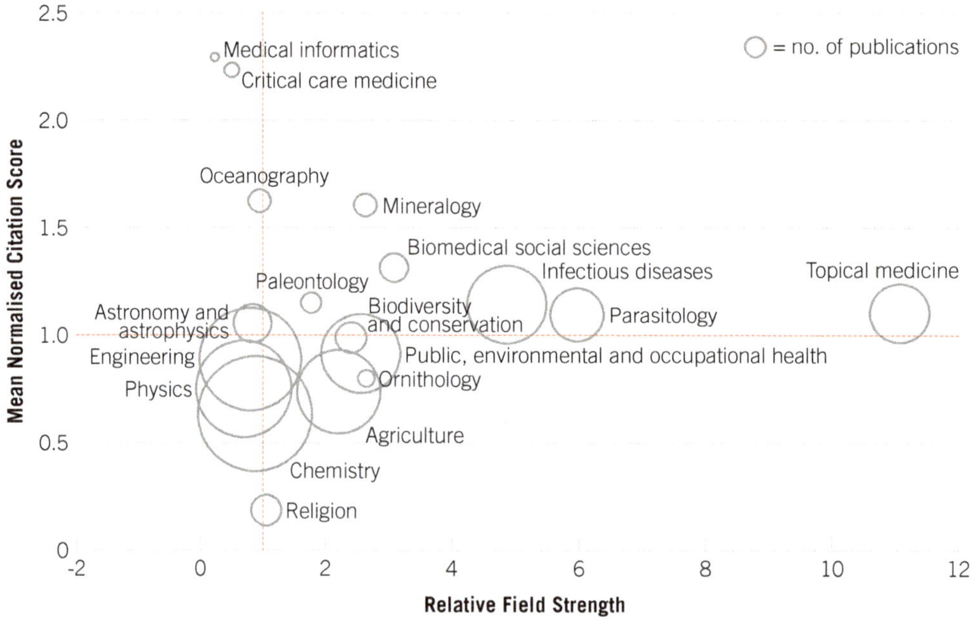

When we plot all the disciplines in the broad field of the agricultural sciences, the results show that two disciplines – veterinary sciences and agricultural engineering – stand out as having above average impact and strength. The field of dairy and animal science is an area in which Africa is relatively strong and its citation impact (just below 1) is noteworthy.

Similar analyses were conducted for other broad scientific fields. In the field of the biological sciences, two fields – reproductive biology, and marine and freshwater biology – emerge as the strongest and high-impact fields. Although Africa is relatively strong in a strategic field such as biotechnology and applied microbiology, its impact on the world stage is very low.

Figure 12: **Positional analysis: Agricultural sciences**

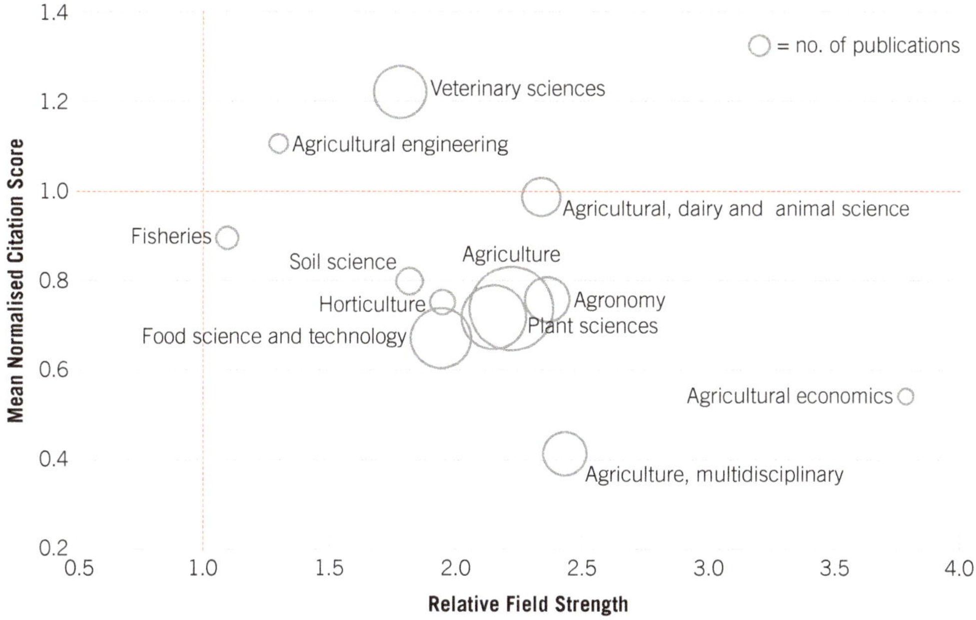

Concluding assessment

This chapter has presented bibliometric evidence in support of the argument that African science has turned the tide in recent years. We have shown how research publications have increased substantially between 2005 and 2015 to the extent that Africa's share of world publications reached 3.2% in 2015. African scientists have increased their research collaborations with the rest of the world and the citation impact of its scientific papers has increased steadily. Our bibliometric analyses of scientific fields reaffirmed Africa's traditional strengths in fields such as tropical medicine, infectious diseases and agriculture.

CHAPTER 3

Research funding landscapes in Africa

Csaba Kozma, Clara Calero Medina and Rodrigo Costas

Introduction

Funding acknowledgement for research funding studies

The impact and influence of funding on scientific publications is a central element of discussion in the scientometric community as well as in the research policy arena (Himanen et al. 2009; Wang & Shapira 2011; Ebadi & Schiffauerova 2016). For research funders, obtaining insights on how their funding schemes are influencing the research landscape is a relevant element for their own development. For other actors in the research landscape, understanding how the different topics and disciplines are the subject of funding programmes is a key issue for the development of their research lines (Ebadi & Schiffauerova 2016; Li et al. 2017).

Bibliometrics provide an excellent platform for the analysis of funding landscapes. This is possible particularly since the Web of Science started to collect 'funding acknowledgement' information from August 2008 onwards. The inclusion of this relevant piece of information has opened new research possibilities in the field of acknowledgements research (Wang & Shapira 2011; Tang et al. 2016) and more specifically in the area of funding acknowledgements studies. The analysis of funding acknowledgements in scientific publications allows for the studying of the presence of funding across disciplines and their evolution over time (Wang & Shapira 2011). Thus, it is possible to obtain global pictures on how the different disciplines are the subject of the different funding schemes and funding programmes.

Funding landscapes in Africa

The influence of funding organisations, besides providing funds for publication, mainly intertwines with the continuity of research programmes in African countries (Beaudry & Mouton 2017). They fulfil the role of donors for research and development projects, and function as the overarching body for within- and intercontinental collaboration (Sonnenwald 2007; OECD 2016; Arvanitis & Mouton 2018). Additionally, they represent an impactful source of employment for local scientists and provide facilities that are potentially inaccessible in their home countries (Beaudry & Mouton 2017; Arvanitis & Mouton 2018). The evaluation of these resources play an essential role by mapping the directions, potentials and impacts that are influenced by the quality and quantity of research funding.

The central objective in this chapter is to describe the main landscape of funding and funders in Africa. To investigate this topic we use the following research questions as guidelines through this chapter: What are the volume and main characteristics of the publications acknowledging a particular source of funding? Do publications with funding acknowledgements exhibit a different level of citation impact as compared to those that do not exhibit such acknowledgements? What are the most important research funders in Africa as a continent, as well as in the most impactful African countries?

Each of these questions will be answered to gain a clear insight to the funding structure and its influence on the scientific contribution of the African continent.

Methodology

Working with funding acknowledgments that originate from the Web of Science is not completely straightforward and it is important to be aware of the main characteristics of the collection and main limitations of these data. A recent paper (Paul-Hus et al. 2016) discusses some of the most relevant caveats that need to be taken into consideration, some of which are presented and discussed in the following methodological description.

Data collection

All African[5] publications covered in the Web of Science (WoS) were collected. For every publication we identified whether it was contributed by authors from African countries only ('Only African') and also whether it was produced through national collaboration only (i.e. all its authors are from the same country). These two additional approaches are incorporated in order to provide more nuanced analyses of the presence and distribution of funding acknowledgements. This is important given the fact that it is not possible to always assign the different funder sources to authors or countries, and thus it is difficult to isolate the funding that comes from African sources or from non-African collaborators. By analysing the publications with only local collaboration (i.e. African or national) it is possible to obtain a snapshot on some of the most important local funders.[6]

Additionally, the following filters were imposed into the dataset in order to obtain a set of publications that allows for an informative analysis of the presence of funding (or the lack thereof) across the different fields and countries:

---

5 The following countries were searched in the 'affiliation' field of the WoS database: Algeria, Angola, Benin, Botswana, Burkina Faso, Burundi, Cameroon, Cape Verde, Central African Republic, Chad, Comoros, Congo, Congo People's Republic, Côte d'Ivoire, Democratic Republic of Congo, Djibouti, Egypt, Equatorial Guinea, Eritrea, Ethiopia, Gabon, Gambia, Ghana, Guinea, Guinea Bissau, Kenya, Lesotho, Liberia, Libya, Madagascar, Malawi, Mali, Mauritania, Mauritius, Morocco, Mozambique, Namibia, Niger, Nigeria, Reunion, Rwanda, Sao Tome & Principe, Senegal, Seychelles, Sierra Leone, Somalia, South Africa, South Sudan, Spanish Sahara, Sudan, Swaziland, Tanzania, Togo, Tunisia, Uganda, Western Sahara, Zambia, Zimbabwe.

6 However, it is important to remark that it is possible that some African funders may be relevant in research fields that involve international (or non-African) collaboration, and as a result they may not be identified in the analysis of the locally collaborative publications.

a. Publications from the period 2009 to 2014 were considered (as indexed in the Centre for Science and Technology Studies (CWTS) in-house database until the week 26 of 2015). Funding acknowledgements (FA) data collection only started in August 2008 (Álvarez-Bornstein et al. 2017; Web of Science 2009). Recently Paul-Hus et al. (2016) observed that, although there is some residual presence of FAs before 2008, the FA data collection before 2009 is not reliable and therefore the most reasonable approach is to focus on the period 2009 to 2014.
b. Only publications covered by the Science Citation Index Expanded (SCIE) database were considered for analysis. It has been observed that for the period 2009–2014 only publications from the SCIE database are actually indexed for funding acknowledgements. Thus, in order to provide a proper picture on the presence and lack of funding, only publications covered by this database were considered for this analysis.
c. Only document types 'article' and 'review' were considered for the analysis, as these are the only document types that have been covered for funding acknowledgements by Web of Science (see Paul-Hus et al. 2016). Similarly, only publications written in English have been considered, as only funding acknowledgements written in English are indexed by Web of Science (Paul-Hus et al. 2016). Thus, in order to restrict our dataset to a homogenous set of publications with potentially only English-language funding acknowledgements, we restrict our analysis to only publications in this language.

Cleaning of funding organisations recorded by Web of Science

The data collected by Clarivate Analytics on the funders mentioned in the acknowledgements of the publications is essentially composed of the raw text strings of the different funding organisations that appear in the funding acknowledgements. This means that funders appear in many different variant forms, thus limiting the usefulness of these data, unless an adequate cleaning and harmonisation of the different variants are performed. In order to identify more accurately the main funders mentioned in the publications, a thesaurus has been developed in order to clean the most important funders present in the whole Web of Science database. This thesaurus follows a similar philosophy and structure as the harmonisation and cleaning of the affiliations performed for the Leiden Ranking (Waltman et al. 2012). CWTS has already identified major funders and classified them in different institutional types: research organisation; funding organisation; governmental institution; and hospitals and funding channels (Van Honk et al. 2016).

An important characteristic of this thesaurus is that it is based on the information contained in the Web of Science as a whole, which means that it is a global thesaurus that is useful when conducting a global analysis of funders. However, it may have some limitations when working at regional or local levels. For this reason, in addition to the most important harmonised variants, other variants (for which it was impossible to perform proper harmonisation) have been also identified, based on their total occurrences in the African dataset. In the case of, for example, the 'National Research Foundation', the references to this funder are not sufficiently clear to determine which 'NRF' funder is actually the one being mentioned in the acknowledgement (e.g. there is a National Research Foundation in South Africa and another one in Singapore).

Analytical approaches and indicators

In order to describe how we conducted our analysis, it is essential to clarify the approaches and exact indicators used. First, the number of publications (P) based on the aforementioned filters (data collection: filters a-c) were counted and those that acknowledge funding (P FA) selected. Based on these data, we calculated the proportion of publications that acknowledge funding (PP FA). In addition, the total citation score (TCS), excluding self-citations, was also calculated. Further, we also calculated the total normalised citation scores (TNCS) and the average number of field-normalised citations of the publications of countries and funding organisations (MNCS), thus correcting for differences stemming from separate scientific fields, publication years and document types. In addition, to select the most cited publications based on their source countries and organisations, we extracted the top 10% of publications that are most frequently cited [P (top 10%)] as well as the proportion of the publications of that, compared with other similar publications, ended up being in the top 10% most frequently cited publications [PP (top 10%)]. We further calculated the impact of the journals in which the selected publications appeared, compared to the world citation average in the subfields covered by these journals (MNJS). Finally, we extracted the proportion of publications that included domestic [PP (collab)] or international collaboration [PP (int collab)].

These indicators pave the way for our analysis to produce a quantified picture of the African funding landscape.

Results

Based on our analysis, we first describe the main African countries that have the highest number of publications (P), highest number of publications that share funding acknowledgements (P FA), and the highest proportion of publications that acknowledge funding (PP FA). As a second step, we present the prominent funders that contribute the most financial support to the publications originating from the African continent. Finally, we depict a thematic funding landscape in which we show the amount of publications as well as the share of publications with funding acknowledgments belonging to certain fields and funding bodies.

The overall analysis of the presence of funding mentions in African publications is presented in Table 4. It shows the presence and proportion of publications that acknowledge funding for the ten main African countries in terms of their overall number of publications. These countries are South Africa, Egypt, Tunisia, Nigeria, Algeria, Morocco, Kenya, Ethiopia, Uganda and Tanzania.

When all African publications are considered (i.e. regardless if they involve international collaboration) the results show that approximately half contain some mention of sources of funding. However, when the focus moves to only African publications (i.e. those publications only involving the participation of African researchers) then this percentage drops to 31.5%, which is very similar to the percentage we find when we focus only on those publications with national collaboration (30.7%). It is important to note that the number of publications that involve African collaboration (i.e. more than one African country) is relatively low, as the difference between the set of African-only publications and national-only publications is very small (1 740 publications) (Beaudry & Mouton 2017). This suggests that most of the African-only collaborations are national collaborations and that the level of collaboration involving only African countries is relatively low (around 2% of African-only publications are actually between more than one African country) (Beaudry & Mouton 2017).

Figure 13: **Presence of publications with funding acknowledgements in Africa (2009–2014)**

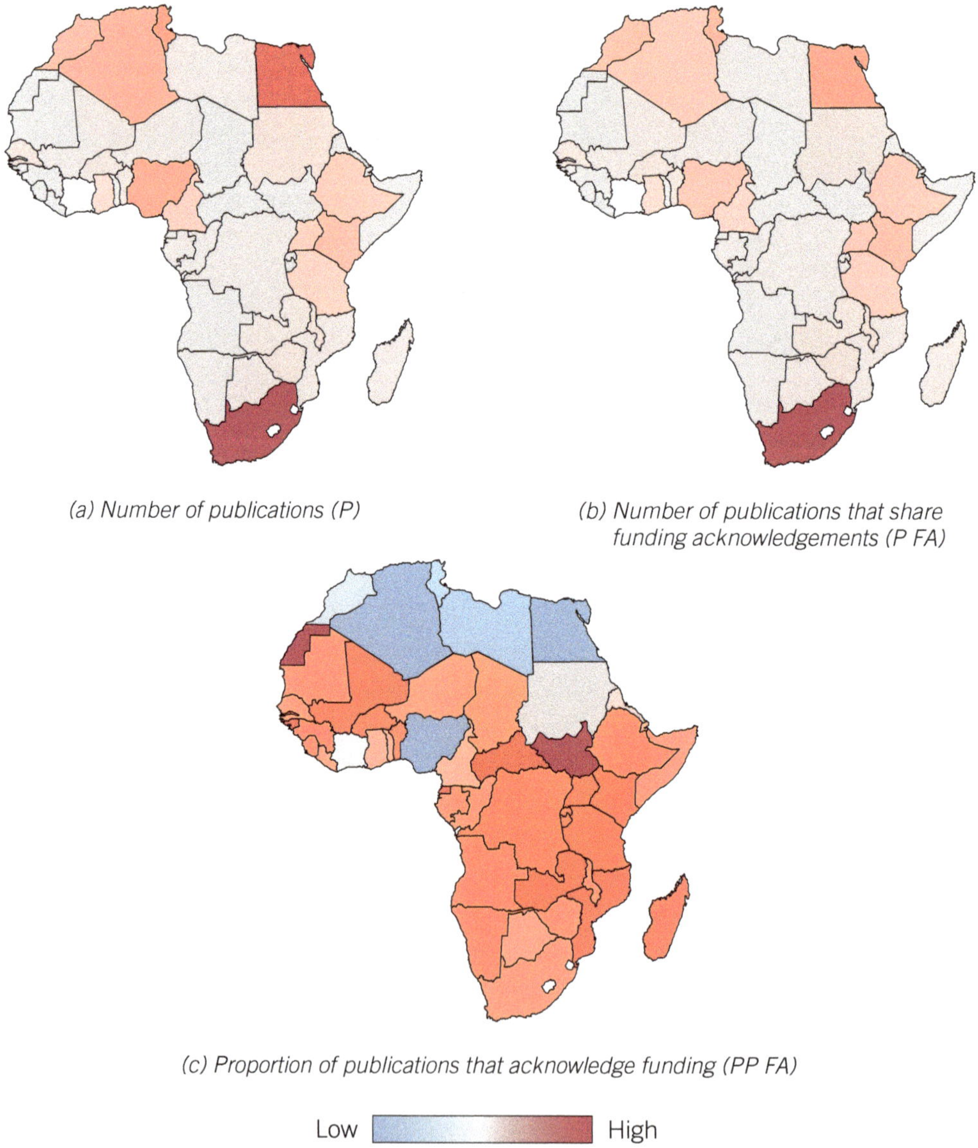

(a) Number of publications (P)

(b) Number of publications that share funding acknowledgements (P FA)

(c) Proportion of publications that acknowledge funding (PP FA)

Note: Warmer colours represent higher numbers

Table 4: **Overall analysis of the presence of funding mentions in African publications between 2009 and 2014 (indicators)**

Area	Publication set	P	% total output	TCS	TNCS	MNCS	P (top 10%)	PP (top 10%)	MNJS	PP (collab)	PP (int collab)
Africa	Total output	170 776	100.0%	902 996	147 690.31	0.86	13 023	7.6%	0.89	73.8%	59.1%
Africa	Funded publications	87 152	51.0%	574 903	91 750.68	1.05	8 491	9.7%	1.05	85.2%	75.3%
Africa	Not funded	83 624	49.0%	328 093	55 939.64	0.67	4 531	5.4%	0.72	62.0%	42.2%
Subset of publications	**Publication set**	**P**	**% total output**	**TCS**	**TNCS**	**MNCS**	**P (top 10%)**	**PP (top 10%)**	**MNJS**	**PP (collab)**	**PP (int collab)**
Only Africa	Total output	71 443	100.0%	238 214	39 929.34	0.56	2 854	4.0%	0.66	39.2%	4.5%
Only Africa	Funded publications	22 479	31.5%	85 535	14 205.24	0.63	1 050	4.7%	0.76	44.9%	7.6%
Only Africa	Not funded	48 964	68.5	152 679	25 724.10	0.53	1 804	3.7%	0.62	36.6%	3.0%
Nat. publications	Total output	69 703	100.0%	235 802	39 386.02	0.57	2 851	4.1%	0.67	36.1%	0.0%
Nat. publications	Funded publications	21 420	30.7%	83 447	13 751.98	0.64	1 028	4.8%	0.76	40.0%	0.0%
Nat. publications	Not funded	48 283	69.3%	152 355	25 634.03	0.53	1 823	3.8%	0.62	34.4%	0.0%
Main African countries	**Publication set**	**P**	**% total output**	**TCS**	**TNCS**	**MNCS**	**P (top 10%)**	**PP (top 10%)**	**MNJS**	**PP (collab)**	**PP (int collab)**
South Africa	Total output	48 259	100.0%	332 928	51 661.62	1.07	4 733.25	9.8%	0.99	71.1%	55.9%
South Africa	Funded publications	32 477	67.3%	255 846	39 436.66	1.21	3 653.76	11.3%	1.09	75.7%	61.8%
South Africa	Not funded	15 782	32.7%	77 082	12 224.96	0.77	1 079.49	6.8%	0.78	61.6%	43.9%
Egypt	Total output	41 284	100.0%	187 289	32 325.46	0.78	2 765.00	6.7%	0.77	66.4%	51.0%
Egypt	Funded publications	12 918	31.3%	75 758	12 973.76	1.00	1 202.71	9.3%	0.94	90.7%	85.3%
Egypt	Not funded	28 366	68.7%	111 531	19 351.70	0.68	1 562.29	5.5%	0.69	55.4%	35.3%
Tunisia	Total output	15 820	100.0%	64 401	11 097.43	0.70	890.70	5.6%	0.79	73.7%	55.8%
Tunisia	Funded publications	6 271	39.6%	30 066	4 882.38	0.78	415.45	6.6%	0.88	84.4%	70.6%
Tunisia	Not funded	9 549	60.4%	34 335	6 215.05	0.65	475.25	5.0%	0.73	66.7%	46.1%

Main African countries	Publication set	P	% total output	TCS	TNCS	MNCS	P (top 10%)	PP (top 10%)	MNJS	PP (collab)	PP (int collab)
Nigeria	Total output	12 223	100.0%	45 558	6 986.29	0.57	457.37	3.7%	0.65	62.0%	37.9%
Nigeria	Funded publications	3 703	30.3%	23 845	3 706.78	1.00	298.76	8.1%	0.95	85.2%	75.5%
Nigeria	Not funded	8 520	69.7%	21 713	3 279.52	0.38	158.61	1.9%	0.51	51.9%	21.6%
Algeria	Total output	11 057	100.0%	39 919	7 913.51	0.72	663.45	6.0%	0.75	75.1%	59.9%
Algeria	Funded publications	3 240	29.3%	14 142	2 527.86	0.78	219.16	6.8%	0.84	87.3%	79.1%
Algeria	Not funded	7 817	70.7%	25 777	5 385.65	0.69	444.30	5.7%	0.71	70.1%	52.0%
Morocco	Total output	7 742	100.0%	37 572	6 638.92	0.86	605.61	7.8%	0.84	83.4%	70.3%
Morocco	Funded publications	3 788	48.9%	24 854	4 291.73	1.13	402.82	10.6%	0.99	95.0%	89.3%
Morocco	Not funded	3 954	51.1%	12 718	2 347.19	0.59	202.79	5.1%	0.70	72.2%	52.1%
Kenya	Total output	7 087	100.0%	62 237	9 623.73	1.36	925.15	13.1%	1.28	92.8%	85.9%
Kenya	Funded publications	5 544	78.2%	52 570	8 061.74	1.45	774.40	14.0%	1.35	95.3%	89.4%
Kenya	Not funded	1 543	21.8%	9 667	1 561.99	1.01	150.75	9.8%	1.04	83.7%	73.2%
Ethiopia	Total output	4 140	100.0%	20 131	3 966.33	0.96	322.01	7.8%	0.97	83.0%	67.8%
Ethiopia	Funded publications	2 939	71.0%	14 491	2 877.57	0.98	241.17	8.2%	1.04	87.2%	72.0%
Ethiopia	Not funded	1 201	29.0%	5 640	1 088.77	0.91	80.84	6.7%	0.80	72.8%	57.7%
Uganda	Total output	3 933	100.0%	36 043	4 949.67	1.26	476.10	12.1%	1.24	92.8%	87.3%
Uganda	Funded publications	3 078	78.3%	30 447	4 184.20	1.36	399.57	13.0%	1.32	95.1%	90.6%
Uganda	Not funded	855	21.7%	5 596	765.47	0.90	76.53	9.0%	0.93	84.6%	75.2%
Tanzania	Total output	3 720	100.0%	30 805	4 580.99	1.23	488.74	13.1%	1.23	93.3%	88.6%
Tanzania	Funded publications	2 953	79.4%	26 218	3 885.19	1.32	416.00	14.1%	1.30	94.6%	90.9%
Tanzania	Not funded	767	20.6%	4 587	695.80	0.91	72.74	9.5%	0.96	88.3%	79.9%

Table 4 also shows the impact and collaboration indicators of African publications, depending on whether they present funding information or not. The most important observation is that, in all cases, publications with funding mentions tend to present a higher average impact (MNCS) as well as higher percentages of top 10% highly cited publications, and they are generally published in journals with higher impact (MNJS). This is confirmed for the overall set of African publications, but also for the set of African-only publications and national publications.

Regarding the levels of collaboration, the results show that publications with funding tend to generally have higher levels of collaboration overall, as well as higher levels of international collaboration. For example, for the overall set of African publications, those mentioning some funding involve, proportionately, almost two times more international collaboration compared to those publications that do not exhibit such a mention, and a much higher level of collaborative work. This is also the case when the focus is on the African-only publications. For the only-national-collaboration publications, obviously the level of international collaboration is zero, but still it is possible to discern the higher level of collaboration (i.e. participation of more than one institution from the same country) in the funded publications.

To further elaborate on these results, we depicted the level and impact of publications, as well as the involvement of domestic or international collaboration, according to the presence or absence of funding acknowledgements (see Figure 14).

Figure 14: **(A) describes the distribution of the countries that produce the top 10% most cited publications between 2009 and 2014. The x axis groups these countries according to the presence or absence of mentioned funding, and the y axis represents the proportion of the actual countries. (B) follows a similar concept, except that, instead of the top 10% most cited publications of countries, the world citation average impact in the subfields (MNCS) is depicted. (C) and (D) show the different funding groups regarding domestic (C) and international collaboration (D).**

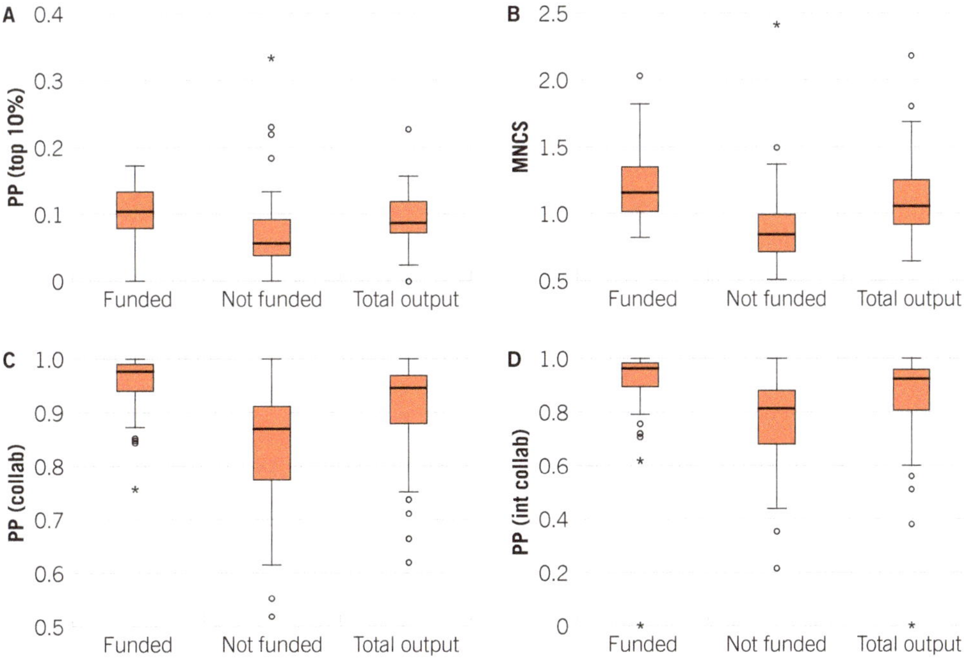

In general, the results show that, between 2009 and 2014, the proportion of publications with funding mentions is higher for the overall set of publications than for those publications involving only local collaboration. This supports the idea that the presence of funding is related to the presence of higher levels of collaboration in scientific publications (cf. Costas & Van Leeuwen 2012; Sonnenwald 2007).

Table 4 also suggests the existence of two main groups of countries. Some countries (such as South Africa, Kenya, Ethiopia, Uganda and Tanzania), in terms of their total and local output, exhibit overall high levels of funding (higher than 50%). North African countries (such as Egypt, Tunisia, Nigeria, Algeria or Morocco) display lower shares of publications with funding (generally lower than 50%). For example, it is remarkable how less than one-third of all Egyptian publications show some funding involvement. In general, the main pattern highlighted here is that, for all countries, publications acknowledging funding exhibit a higher impact and collaboration levels when compared to those publications without explicit funding mentioning.

Prominent funders in Africa

In this section the focus is on highlighting and analysing the most important funders that can be identified through the funding acknowledgements recorded by Web of Science in publications originating from the African continent.

Table 5 lists the funders that, between 2009 and 2014, have been associated with more than 1 220 publications in the overall African set of publications. In the case of Africa, one may assume that most of the 'National Research Foundation' publications very likely belong to the South African NRF, therefore in Table 5 they are presented combined. Apart from the predominant role of the NRF, it is also possible to point out the relevance of some important international funders. Thus, we can mention the European Union, followed by the National Institutes of Health (NIH) of the USA, the Wellcome Trust, and the German Deutsche Forschungsgemeinschaft (DFG). A substantial number of publications also received funding from a diversity of companies ('Companies*'). Other funders that have a relevant influence in African research are the Bill & Melinda Gates Foundation, the Government of Spain, the National Natural Science Foundation of China, the Ministry of Higher Education Scientific Research of Tunisia, and the National Institute of Allergy and Infectious Diseases (which is part of the NIH).

Table 5: **Prominent funding organisations in Africa (output > 1 220 pubs between 2009 and 2014)**

Funder	Funder country	P	TCS	TNCS	MNCS	P (top 10%)	PP (top 10%)	MNJS	PP (collab)	PP (int collab)
National Research Foundation	South Africa	11 726	66 526	10 924.89	0.93	944.83	8.1%	0.94	62.9%	45.4%
European Union	Europe	3 734	44 404	7 785.17	2.08	707.63	19.0%	1.41	96.7%	94.2%
National Institutes of Health	USA	3 072	52 248	5 951.92	1.94	601.59	19.6%	1.70	98.7%	97.1%
Wellcome Trust	UK	2 663	39 678	5 047.49	1.90	560.23	21.0%	1.77	97.4%	94.6%
Deutsche Forschungsgemeinschaft	Germany	2 154	26 457	4 319.05	2.01	403.13	18.7%	1.35	98.2%	97.6%
Companies*		2 045	44 208	5 136.34	2.51	492.37	24.1%	1.74	96.0%	90.8%
Bill & Melinda Gates Foundation	USA	1 963	36 307	4 923.49	2.51	455.11	23.2%	1.92	97.5%	96.0%
Government of Spain	Spain	1 950	27 973	4 561.99	2.34	376.04	19.3%	1.37	99.5%	99.2%
National Natural Science Foundation of China	China	1 887	21 834	4 099.38	2.17	387.18	20.5%	1.27	99.8%	99.7%
Ministry of Higher Education and Scientific Research	Tunisia	1 729	6 911	1 174.48	0.68	90.98	5.3%	0.76	75.2%	54.0%
National Institute of Allergy and Infectious Diseases	USA	1 580	22 158	2 906.89	1.84	310.08	19.6%	1.69	99.1%	96.9%
Federal Ministry of Education and Research	Germany	1 357	22 480	3 549.95	2.62	328.59	24.2%	1.46	99.1%	98.1%
National Science Foundation	USA	1 318	22 864	3 788.95	2.87	363.66	27.6%	1.57	99.2%	99.0%
King Saud University	Saudi Arabia	1 316	2 856	664.35	0.50	48.29	3.7%	0.55	99.0%	98.6%
Medical Research Council	South Africa	1 298	10 869	1 316.28	1.01	133.61	10.3%	1.13	69.8%	47.7%
European Commission	Europe	1 290	13 184	1 909.24	1.48	206.51	16.0%	1.36	97.6%	95.1%
German Academic Exchange Service	Germany	1 278	5 916	997.43	0.78	86.21	6.7%	0.93	86.5%	78.3%
Natural Sciences and Engineering Research Council of Canada	Canada	1 259	14 313	2 335.69	1.86	218.20	17.3%	1.29	99.0%	98.5%
United States Agency for International Development	USA	1 225	13 273	1 800.77	1.47	195.74	16.0%	1.38	94.4%	88.2%

In terms of impact, it seems that most of the publications funded by the main funders exhibit a high citation impact. Particularly relevant is the impact of publications funded by organisations such as the Bill & Melinda Gates Foundation or the set of publications funded by private companies. The National Research Foundation (NRF) from South Africa funded publications have generally achieved an average impact approximating the international level (i.e. MNCS=0.93 or MNJS=0.94).

Regarding collaboration, all non-African funders are mentioned in publications with high levels of collaboration, both overall and international. These levels are lower for the two main African funders – the South African NRF and the Tunisian Ministry of Higher Education and Scientific Research – although in both cases approximately half of the publications funded by these two organisations involve international collaboration.

Thematic funding landscapes in Africa

Capitalising on the results described above, in this section, we focus on the presence of funding across disciplines. Thus, the main approach is to highlight areas with higher or lower levels of funding and how the proportion of publications mentioning any funding has evolved over time. The landscapes are presented in the general map of Web of Science subject categories, excluding those subject categories that do not belong to the Science Citation Index Expanded (SCIE) (i.e. social sciences and arts & humanities subject categories, which are mostly located in the left-hand side of the map).

In Figures 15 and 16, the main funding landscape for all African and world publications, between 2009 and 2014, is presented via the proportion of publications that acknowledge funding (PP FA). The size of the nodes indicates the number of publications in Africa in a given SCIE subject category, while the colour indicates the amount of the share of publications with funding acknowledgments in the field. Nodes with more red colours point to disciplines with higher shares of publications with funding mentions, while blue colours indicate the contrary pattern.

It is visible based on the depiction in Figures 15 and 16 that important fields in African scientific production in terms of higher shares of publications with funding include among others, 'physics, particles & fields', 'ecology', 'astronomy & astrophysics', 'genetics & heredity', 'immunology', 'infectious diseases' and 'virology'.

Funding acknowledgment activity of main funders in Africa

Figures 17 to 22 show the thematic distribution of funding based on the top funding organisations that were mentioned in Table 5. The colour coding refers to the proportion of publications that are funded by the specific organisation compared to all of the funded publications across scientific fields. In each of the figures fields in which the funder appears in more than 10% of the funded publications for the continent are coloured in red. Thus, it is possible to see how the South African NRF plays an important role in most of the natural science fields on the continent, with most of the fields on the right-hand side of the map with 10% of funding by the NRF (Figure 17). The European Union (Figure 18) funding has a strong importance in the funding of astronomy and astrophysics research, with a similar pattern as for the German DFG (Figure 19), while the NIH (Figure 20), the Wellcome

Trust (Figure 21) and the Bill & Melinda Gates Foundation (Figure 22) have a stronger focus on funding health-related research, particularly in the fields of infectious diseases and immunology.

Figure 15: **Main funding landscape for all African publications, between 2009 and 2014, via the proportion of publications that acknowledge funding (PP FA)**

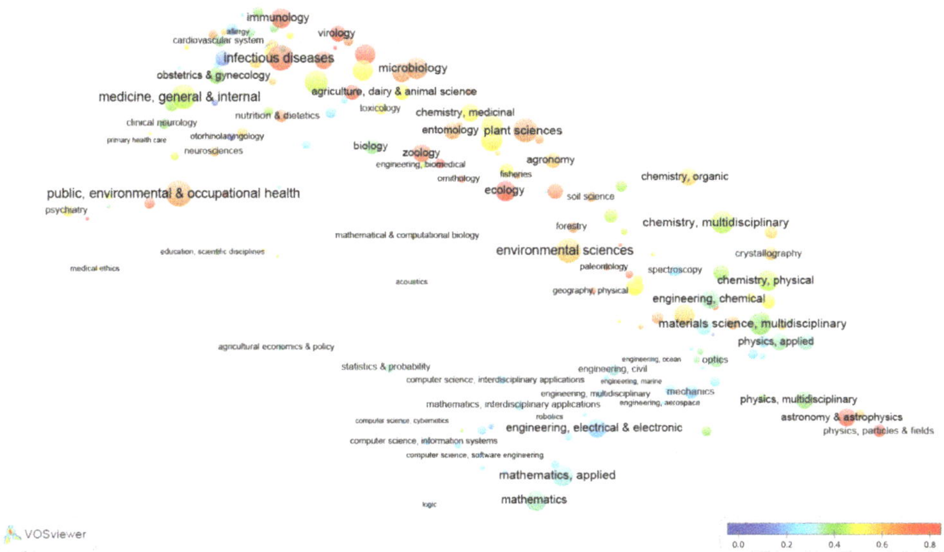

Figure 16: **Main funding landscape for all the world publications, between 2009 and 2014, via the proportion of publications that acknowledge funding (PP FA)**

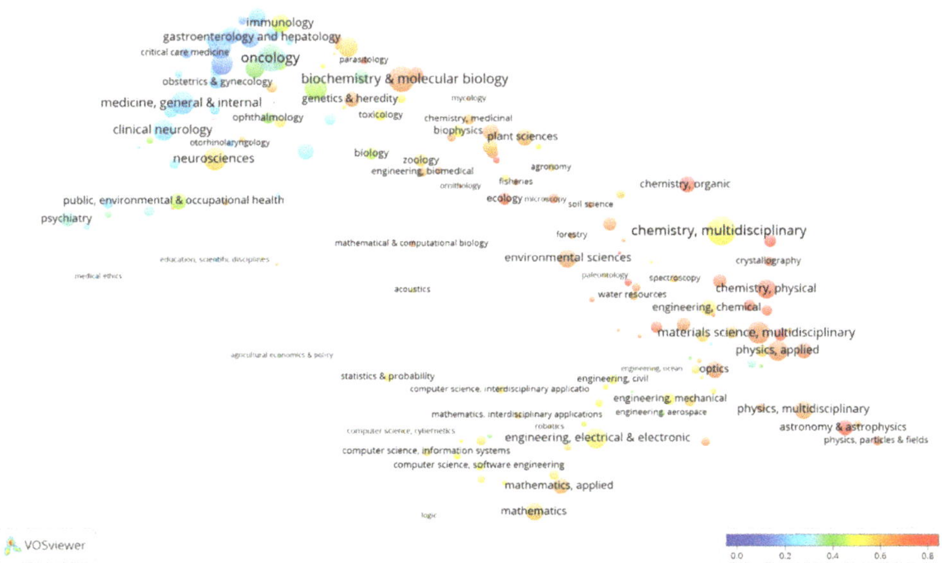

Figure 17: Main funding landscape the South African NRF, between 2009 and 2014, proportion of publications that acknowledge funding from NRF (for all funded publications in Africa)

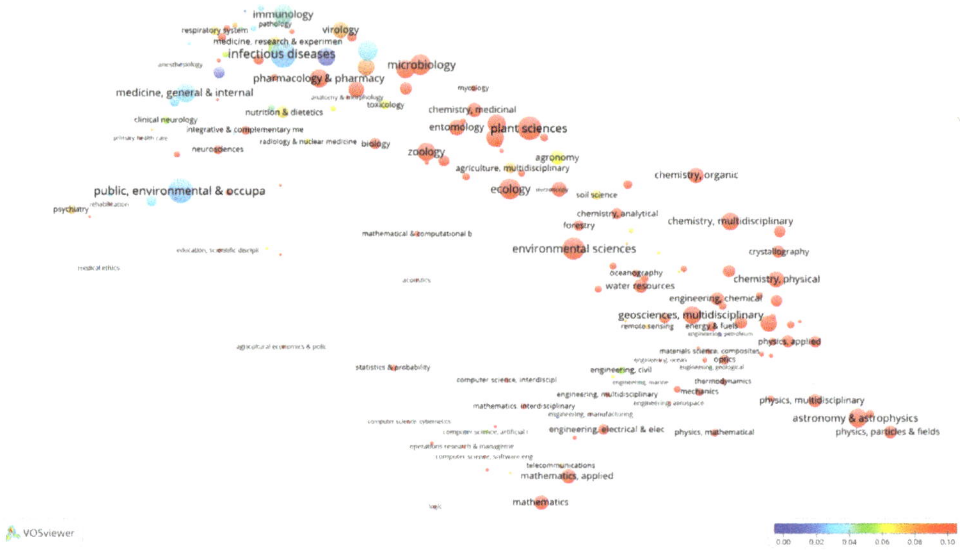

Figure 18: Main funding landscape the European Union in Africa, between 2009 and 2014, proportion of publications that acknowledge funding from the EU (for all funded publications in Africa)

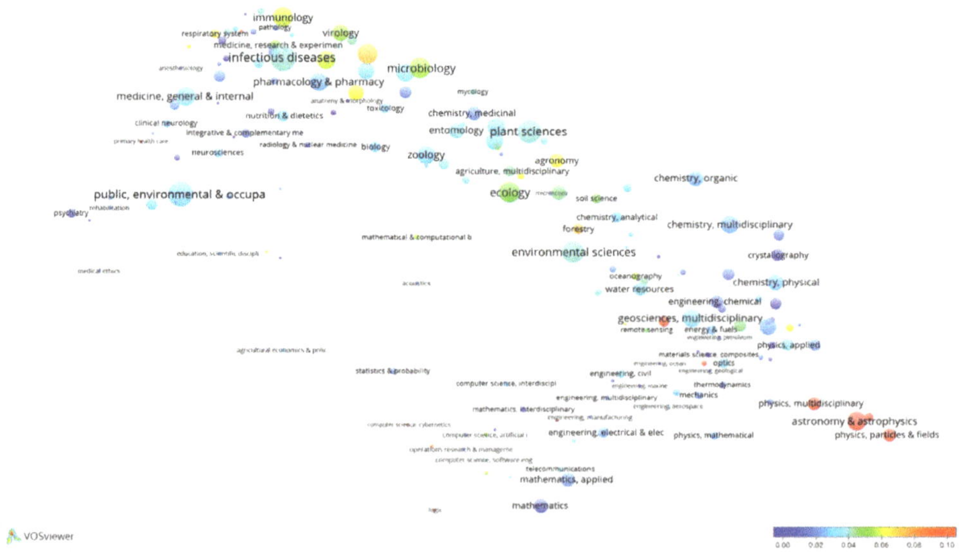

Figure 19: Main funding landscape the NIH in Africa, between 2009 and 2014, proportion of publications that acknowledge funding from the NIH (for all funded publications in Africa)

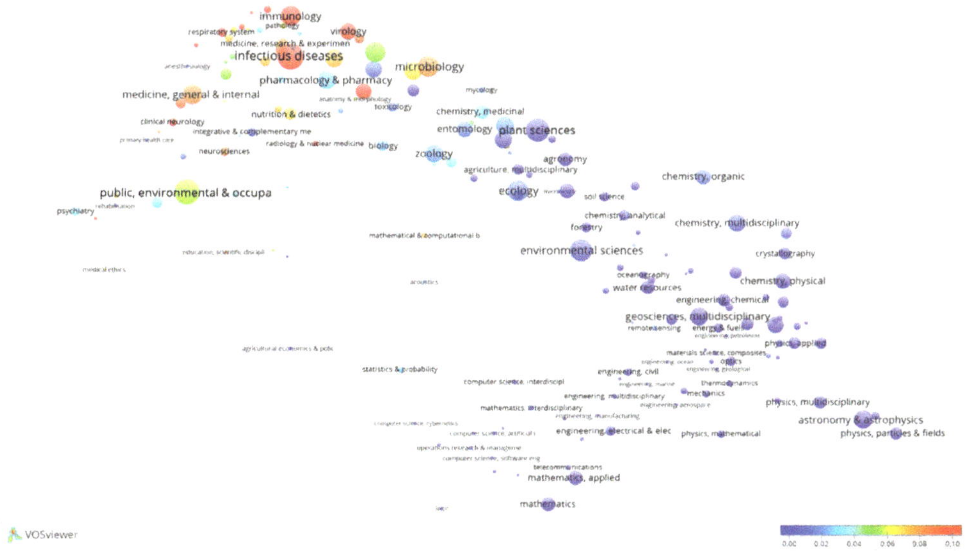

Figure 20: Main funding landscape the Wellcome Trust in Africa, between 2009 and 2014, proportion of publications that acknowledge funding from the Wellcome Trust (for all funded publications in Africa)

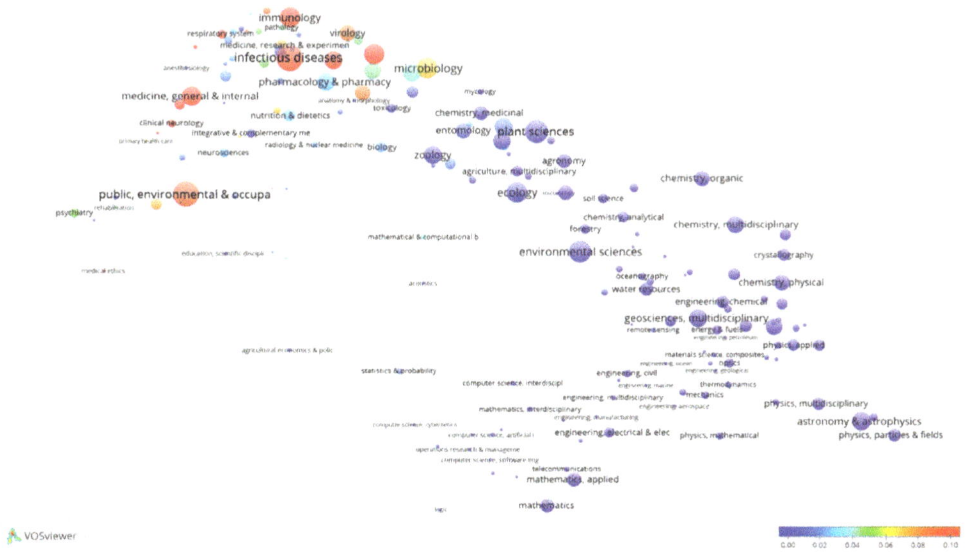

Figure 21: **Main funding landscape the Deutsche Forschungsgemeinschaft in Africa, between 2009 and 2014, proportion of publications that acknowledge funding from the DFG (for all funded publications in Africa)**

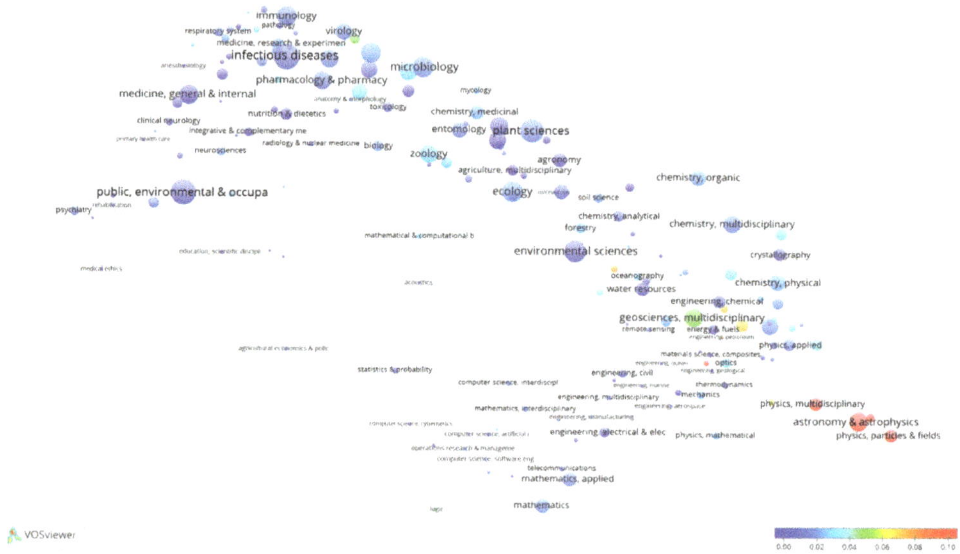

Figure 22: **Main funding landscape the Bill & Melinda Gates Foundation in Africa, between 2009 and 2014, proportion of publications that acknowledge funding from the Bill & Melinda Gates Foundation (for all funded publications in Africa)**

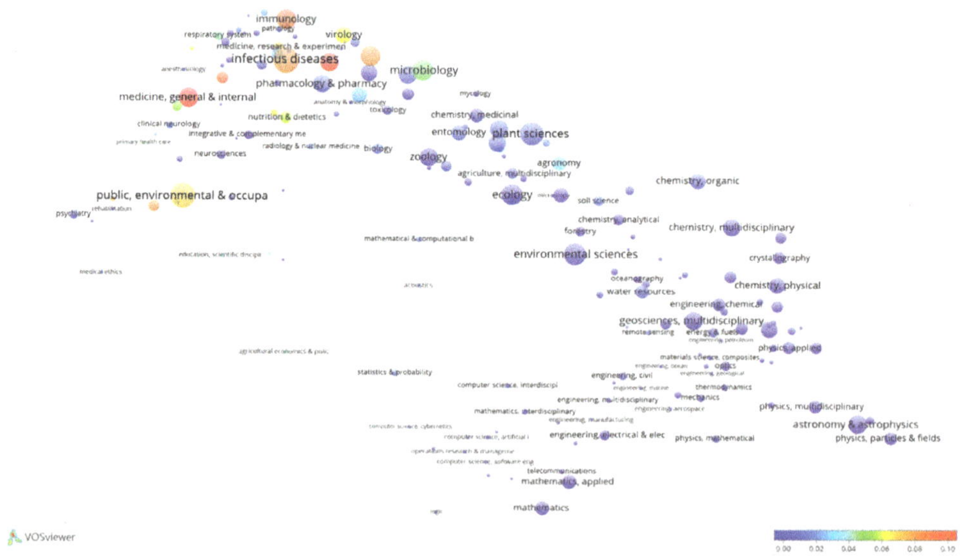

Discussion and conclusions

The inclusion of funding acknowledgments in the Web of Science records since August 2008 has opened up a series of essential and informative analytical solutions in scientometrics. Some previous approaches have been already proposed, including studies by Shapira and Wang (2010) and Wang and Shapira (2011) about the impact of funding regarding the directions and development of research (in this case in nanotechnology) via detailed analysis of funding acknowledgment. In addition, prior critical evaluations of general funding acknowledgments attributes have been proposed by Rigby (2011) who highlighted the simple errors (erroneous inclusion of names of funding bodies) and lack of inclusion of political and cultural influence of funding acknowledgment structure that stem mainly from the generally poor standardisation of funding information regarding funding acknowledgment.

Considering the previous approaches, this chapter describes a general analysis of the presence of funding acknowledgments in African publications and across research fields. The main idea was to introduce a general landscape on the presence, distribution and impact of publications that acknowledge some type of funding support. Also, a discussion on the presence and increase of funding information in scientific publications across scientific fields has been performed, thus showing the disciplinary landscape of African publications in terms of the presence (or lack) of funding acknowledgments.

The important element to take into consideration is that funding acknowledgment data are not free of limitations (cf. Paul-Hus et al. 2016) and it is essential to consider these restrictions in the interpretation of the results based on these data. Thus, the results presented here need to be considered in the perspective of the following limitations:

- This funding acknowledgment analysis relies on the methodology used by Clarivate Analytics to identify, extract and parse the funding acknowledgment information present in scientific publications. The possibility of errors, omissions and inconsistencies in this process introduces the need of care when working with funding acknowledgements data.
- Potential non-disclosed changes in the funding acknowledgment indexing policies of Clarivate Analytics need to be considered (either over time or regarding the different sources considered for indexing), call for care in the interpretation and use of funding acknowledgment data.

Our results show that 51% of all African publications exhibit some mention of funding sources. However, this share decreases to 32% when the focus is only on publication carried out by African authors. This can be related to the fact that publications with funding acknowledgements tend to be collaborative publications, thus publications done only by African authors involve lower levels of collaboration (both overall collaboration and international collaboration, which is confirmed in Table 2) which also exhibit lower levels of funding acknowledgements. The higher collaborative nature of funded publications has been already discussed (Costas & Van Leeuwen 2012; Defazio et al. 2009; Sonnenwald 2007), pointing to the close relationships that these two concepts have, and suggesting that funding may play an important role in setting and facilitating collaborations among scholars and countries.

In terms of citation impact, funded publications exhibit a higher average field-normalised impact as compared to those publications that do not mention any funding source. Also, they present higher shares of top 10% highly cited papers and are published in journals of much higher impact. These results are in line with previous studies that have already pointed to the higher citation impact of publications with funding acknowledgements (Giles & Councill 2004; Levitt 2011; Costas & Van Leeuwen 2012).

The analysis focusing on the most productive countries suggests the existence of two main groups of countries. A group composed by South Africa, Kenya, Ethiopia, Uganda and Tanzania with high shares of funding mentions (above 70% for their whole set of publications) and a second group composed by countries like Egypt, Tunisia, Nigeria, Algeria and Morocco with shares of funding acknowledgments below 50%. These differences could be attributed to the presence of strong national funding agencies (e.g. the National Research Foundation in South Africa) or foreign funders (such as the Wellcome Trust, the NIH or the Bill & Melinda Gates Foundation in the cases of Kenya, Ethiopia, Uganda and Tanzania).

For countries like Egypt or Morocco, their local funders (e.g. the Ministry of Higher Education of Egypt or the *Centre National pour la Recherche Scientifique et Technique* in Morocco) are at the same level of activity as other foreign funders. In Tunisia, the strongest funder is the Ministry of Higher Education and Scientific Research, but still with a lower presence in the output of the country. For countries like Nigeria or Algeria there are no strong local or foreign funders that play a strong role in their scientific production. All in all, the whole production of the continent is characterised by the presence of non-African funders, with the European Union, the NIH and the Wellcome Trust as some of the most important examples. From an African perspective, the role of South African funders (e.g. the NRF, the South African Medical Research Council, or South African universities) is predominant in the continent, not highlighting other major funders from other countries; putting aside the case of the Ministry of Higher Education and Scientific Research of Tunisia.

The funding landscapes show some important biomedical areas exhibiting high levels of funding presence, including disciplines like 'infectious diseases', 'immunology', 'virology', 'parasitology', 'tropical medicine' and 'health care sciences & services' with funding acknowledgement shares higher than 70%. There is also an important presence of funding around areas related with 'ecology', 'zoology', 'evolutionary biology' and 'biodiversity conservation' (also with shares higher than 70%). Other disciplines that also present high levels of funding presence include 'astronomy & astrophysics', 'physics, particles & fields' and 'genetics & heredity'. When the focus moves toward Africa-only production, the same fields are funded, although some other fields such as 'marine & freshwater biology' and 'forestry' also emerge as having strong local-level of funding.

In terms of growing levels of shares of funding mentions in publications, three main areas can be highlighted, one related with 'computer sciences', 'mathematics' and 'engineering'; a second one related with natural sciences fields, mostly on 'chemistry' and 'physics'; and third fields related with 'general and internal medicine' and 'infectious diseases'. Disciplines from these areas exhibit highest increases, although the general picture is that the share of funding mentions is increasing in most of the disciplines.

In conclusion, we can see that the presence of funding acknowledgements has an important relationship with numerous essential indicators of quantitative scientific activity across the African continent, which urges the more widespread and standardised appearance of this feature in future publications.

PART TWO

The challenges that young scientists in Africa face

CHAPTER 4

The young scientist: A profile

Catherine Beaudry, Johann Mouton and Heidi Prozesky

Introduction

This study is about the young scientists and scholars in Africa – the next generation of African scientists. We initiated the study in order to gain a better understanding of the careers of young scientists, in particular their career and research decisions, their experiences as scientists, as well as their concerns and the challenges they face in establishing themselves as expert and internationally competitive scholars. In this section of the book we present the findings of our empirical work involving a continent-wide survey as well as a large number of in-depth interviews. As we indicated previously, this study arguably comprised of the largest survey ever conducted amongst scientists, and specifically young scientists, on the African continent. A web-based survey was conducted between May 2016 and February 2017. More than 120 000 questionnaires were distributed through two online survey platforms. When the survey was closed, a total of 7 513 completed questionnaires had been received. This was followed by more than 250 individual, in-depth interviews which were conducted between April 2017 and March 2018.[7] In the following chapters we present the main findings of our research. We begin in this chapter with a sketch of the profile of the young scientists in Africa.

Defining the young scientist

Even a superficial reading of the literature in our area of interest shows that the term 'young scientist' is a contested term. In various studies reference is made to 'early-career academics or scholars', 'emerging scholars' or the 'next generation of scholars'. In the first Global State of Young Scientists Project (GloSYS) study, the definition of 'early career researchers' was the following: 'A Young Scientist is defined as a postgraduate or early career researcher of any discipline actively pursuing a research career, usually without being fully established yet. She/He will have received a PhD or an equivalent doctoral qualification up to 10 years ago and is usually between 30 and 40 years old' (Friesenhahn & Beaudry 2014: 57).

In our initial engagements we decided to use this definition. We expected that a cut-off age of 40 years would be the threshold between early-career and established scientists, and could be employed as such in our study as well. In addition to the criterion of age, the 'operational definition' of a young scientist as presented above, also requires that the

7 More detail about the research design and methodology is presented in Appendix 1.

scientist must be in possession of a PhD or doctorate. When combining these two criteria (the actual age of respondents at the time of the survey as well as their age at the time of receiving a PhD), we found distributions as shown in Figure 23. The mean age is 45 for all the respondents.

Figure 23: **Chronological age and age at PhD of respondents compared**

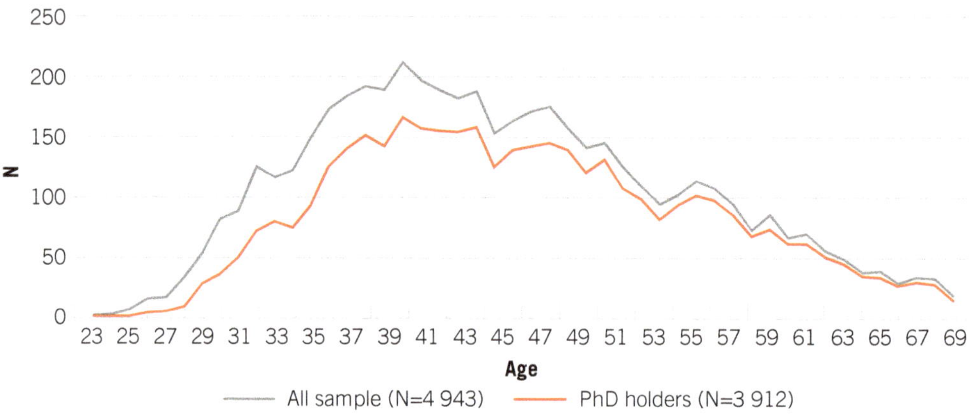

These distributions would suggest that a cut-off criterion of 45 would be appropriate – both for our total sample and for those with PhDs. However, the third criterion mentioned in the GloSYS definition (having received the doctorate up to 10 years previously), does not hold for our sample. Indeed, more than half (54%) of those who graduated within the last 10 years are older than 40.

Figure 24: **Age distribution of respondents who obtained their PhD in the 10 years preceding the survey**

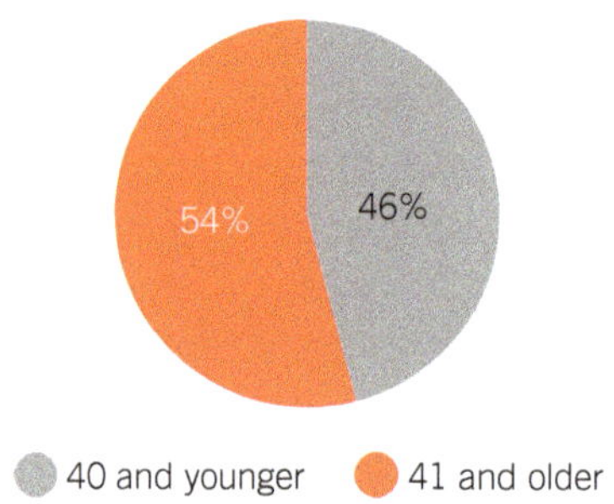

Figure 25: **Age distribution of respondents who published their first research article in a peer-reviewed journal during the 10 years preceding the survey**

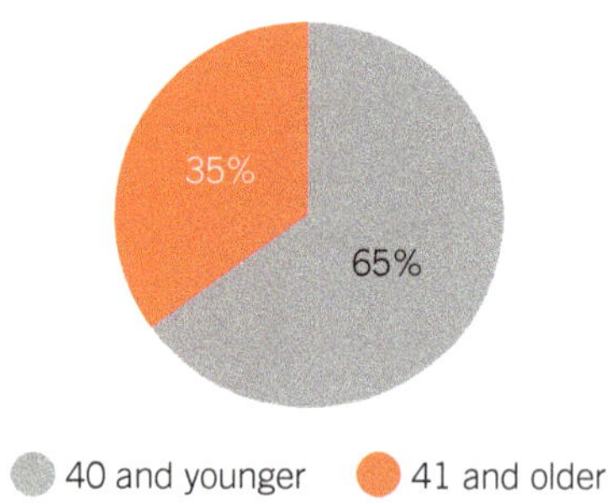

Taking the age at publication of the first article into account yields a similar problem. More than a third of those who published their first research article within the 10 years prior to the survey are older than 40.

With this in mind, we decided to split the cases into several categories using the dispersion of the respondents according to chronological age, age at highest qualification and age when their first article was published. As expected, respondents generally publish their first article at a younger age than when they graduate. PhD students are often expected to publish during their studies, and if a researcher with a master's degree decides to undertake PhD studies, he/she would probably have published before graduating. Since there is a clear correlation between chronological age on the one hand, and age when the highest qualification was obtained and age at the publication of a respondent's first article, on the other, we decided to use only the chronological age of respondents to analyse differences between age groups.

Table 6: **General statistics for chronological age, age at highest qualification, and age at publication of first article**

	Chronological age	Age at highest qualification	Age at first article
N	4 906	4 867	4 322
Mean	45.81	36.05	32.02
Median			
Mode	40	31	29
Std. Dev.	9.829	7.081	6.536
p25	38	31	28
p33	40	32	29
p50	45	35	32
p66	49	38	35
p75	53	40	36

For this purpose we 'tested' two different age classifications. The first (binary) separated our sample into two age categories: 45 and younger and older than 45. The second (ternary) grouped the sample into three age categories: 39 or younger, between 40 and 50, and older than 50. The results of various tests (both CHAID and multiple analysis of variance) consistently showed that the ternary classification into three age categories produced more significant and useful results. We therefore decided to use the ternary classification in all our subsequent thematic analyses.

Gender

The study also has a particular concern with gender and how it correlates with various areas of interest in our study. We set out, as stated in the original problem statement of the study, to specifically establish whether gender correlates with the age, scientific field and other biographical characteristics of our respondents. We also undertook to investigate whether gender correlates with specific issues regarding the research performance, mobility, collaboration and networking behavior of the survey respondents, as well as the reported barriers to their careers as scientists. Hence, in most of our analyses we also disaggregated the results by male and female.

We were not surprised to find that male respondents were in the majority in our sample: 70% male compared to 30% female respondents. This distribution is consistent with data gathered by the UNESCO Institute for Statistics (UIS), which show that men are generally over-represented among the researcher work-force in Africa. Statistics from the UIS shows, however, that the distributions between male and female researchers differ from country to country. For some of the countries we surveyed (Cameroon, Kenya, Morocco, Nigeria, and Tunisia), the gender distributions also differ substantially from those provided by the UIS. For South Africa and Ghana, the percentages of women amongst our survey respondents is almost the same as those reported by the UIS; and for Burkina Faso, Uganda, Zambia and Zimbabwe, the difference is not significant. Disaggregation of the gender of our respondents by the threefold age classification showed no significant differences (Figure 26).

Figure 26: **Gender distribution, by age**

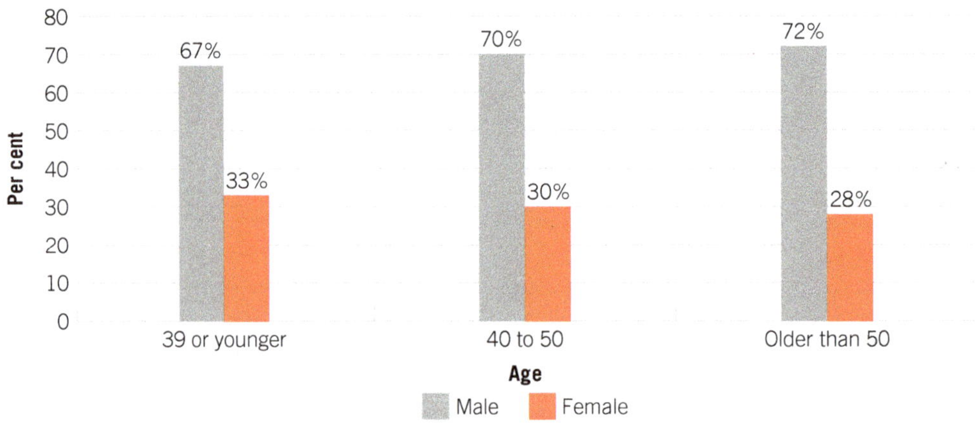

Scientific field

Given the emphasis of our study on the factors that impact on the research performance of young scientists, we decided to add the 'scientific field' of the respondents to all analyses. It is a well-known fact that publication behaviour in particular is highly correlated with scientific field. Scientists and scholars from different fields have different preferences as far as publication strategies are concerned. Whereas most scientists in the natural and health sciences prefer to publish their results in journal articles, many scholars in the humanities and social sciences prioritise books and chapters in books as their preferred publication format. Scientists in such fields as engineering, information sciences and computer science again have a preference for conference proceedings. So the inclusion of scientific field in our analyses is important. But we also wanted to establish whether other research-related practices (such as securing funding for research, collaboration and networking) are field-dependent. Hence we coded the reported discipline in which the respondent received their highest degree into six 'scientific fields' (natural sciences; agricultural sciences; engineering; medical and health sciences; social sciences; and humanities) and included this classification into all statistical analyses (Figure 27). The field profile of the young scientists (39 years or younger) shows that we have slightly higher proportions of respondents in the natural sciences and slightly lower proportions of respondents from the agricultural sciences, when compared to the older cohorts. In general, though, our sub-sample of young scientists do not differ much from the total sample in terms of field profile.

Figure 27: **Scientific field distribution, by age**

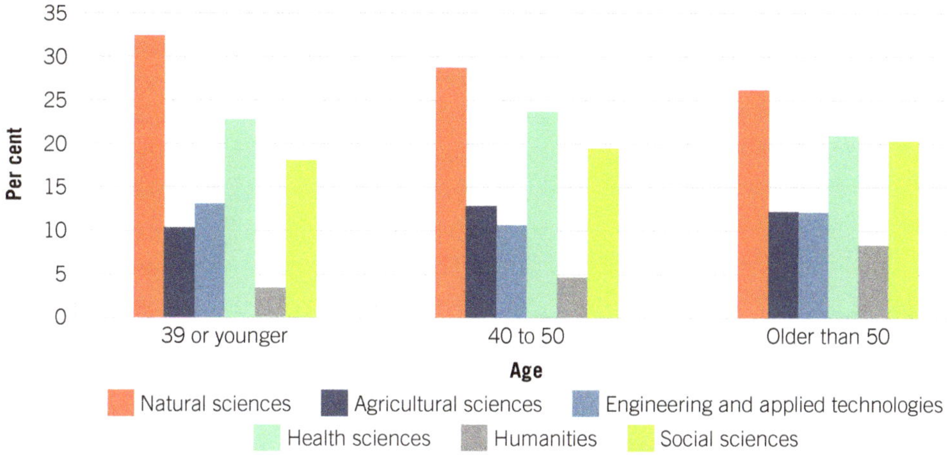

Residence and nationality

Almost 40% of the survey respondents reside in southern Africa, close to another 30% live in West Africa, 12% in East Africa, and 22% in North Africa (see Figure 28).

Figure 28: **Region of residence**

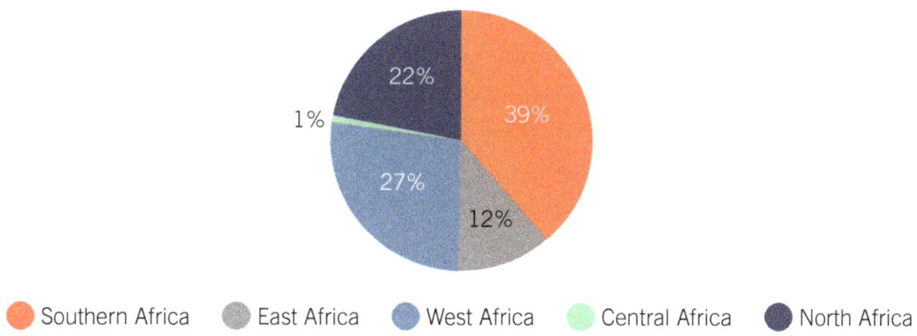

The distribution of the younger scientists is quite similar to that of the population as a whole, as Figure 29 shows.

Figure 29: **Region of residence, by age**

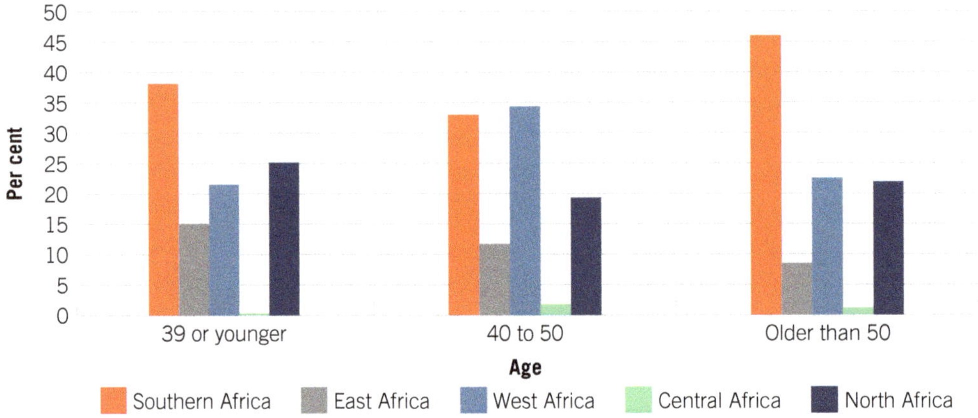

Since approximately 9% of respondents do not live in the same country as their country of nationality, it is also important to consider respondents' nationalities. Table 7 displays respondents' nationalities arranged into the same regions used to categorise country of residence.

Table 7: **Nationalities**

Nationality	N	%	Nationality	N	%
Southern Africa			**East Africa**		
Angola	1	0%	Burundi	3	0%
Botswana	39	1%	Comoros	1	0%
Lesotho	4	0%	Eritrea	3	0%
Malawi	39	1%	Ethiopia	148	3%
Mozambique	2	0%	Kenya	189	4%
Namibia	2	0%	Madagascar	22	0%
South Africa	1 495	30%	Mauritius	4	0%
Swaziland	7	0%	Rwanda	3	0%
Zambia	94	2%	Seychelles	2	0%
Zimbabwe	173	3%	Somalia	1	0%
Total southern African	**1 856**	**37%**	Sudan	1	0%
West Africa			Tanzania	75	1%
Benin	49	1%	Uganda	150	3%
Burkina Faso	63	1%	**Total East African**	**602**	**12%**
Cameroon	180	4%	**Central Africa**		
Chad	2	0%	Central African Republic	5	0%
Côte d'Ivoire	72	1%	Congo – DRC	45	1%
Ghana	125	2%	Congo – Republic	5	0%
Guinea	5	0%	Gabon	12	0%
Mali	24	0%	**Total Central African**	**67**	**1%**

Nationality	N	%	Nationality	N	%
Mauritania	1	0%	**North Africa**		
Niger	16	0%	Algeria	411	8%
Nigeria	765	15%	Egypt	156	3%
Senegal	72	1%	Morocco	227	5%
Sierra Leone	3	0%	Tunisia	308	6%
Togo	33	1%	**Total North African**	**1 102**	**22%**
Total West African	**1 410**	**28%**			

These data have been further aggregated to produce Figure 30 which allows for a comparison of the distributions of across region of residence and region of nationality, which we find to be very similar, irrespective of age (Figure 31).

Figure 30: **Region of nationality**

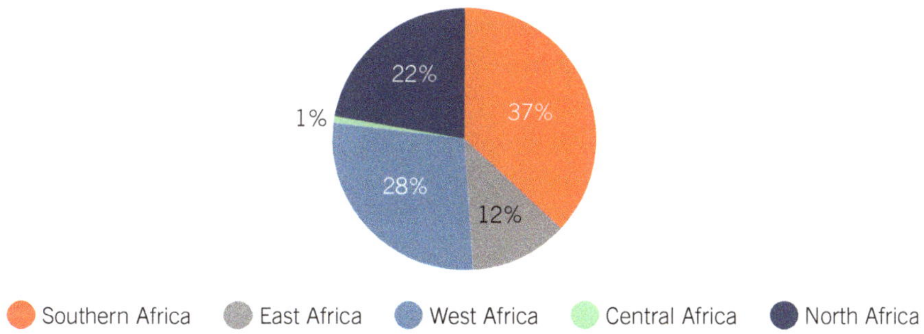

Figure 31: **Region of nationality, by age**

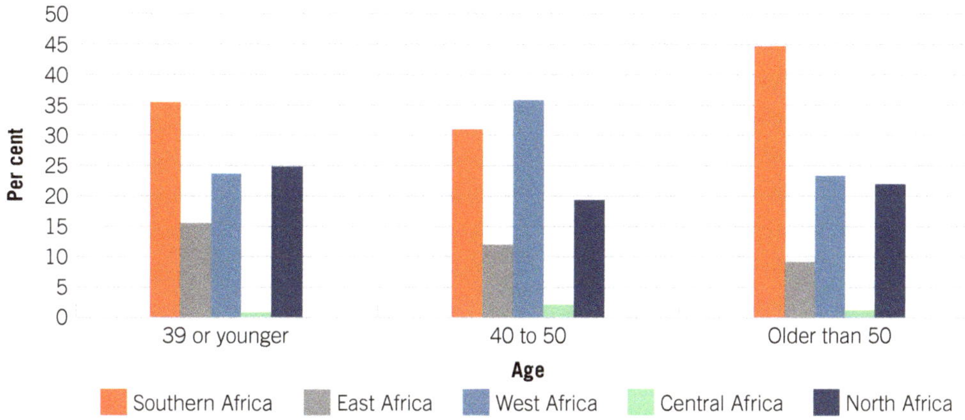

Educational background

Almost all respondents (98%) hold either a PhD (or an equivalent degree) or a master's (or an equivalent) degree. Four out of five respondents are PhD graduates. When disaggregated by age (Figure 32) we found a predictable pattern: Among the young scientists, a sizeable proportion (35%) are not in possession of a doctoral degree.

Figure 32: **Highest qualification, by age**

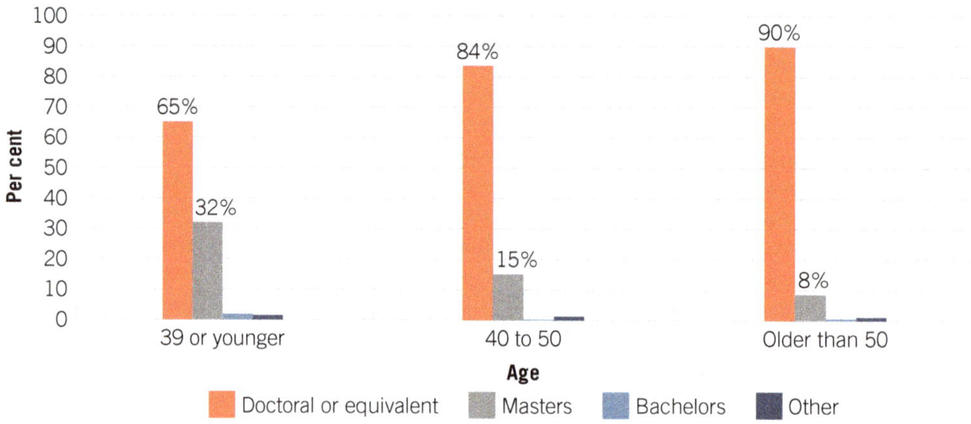

Employment characteristics

Employment sector

Three-quarters of our respondents (76%) are primarily employed in the higher education sector. The second-most frequently indicated employer (by 13% of respondents) is a public research institution, followed by small percentages employed in business enterprises (3%), non-governmental or non-profit organisations (3%), and government institutions (2%). The remaining 3% are distributed amongst private research institutions, private practice, international (research) organisations, and the school sector. Although difficult to prove conclusively, we believe that this distribution basically reflects the concentrations of knowledge production in Africa. The vast majority of research studies and scientific papers are produced at universities on the continent, followed by public research institutes. There are few instances of R&D being performed by private companies or even government departments. Table 8 below disaggregates, again, sector of employment by age category. There are small differences across categories – none of which are statistically significant.

Table 8: **Employment sector, by age**

	39 or younger	40–50	Older than 50	Total
Higher education	68.9%	76.0%	80.7%	75.4%
Public research institution	14.8%	13.3%	10.3%	12.8%
Business enterprise	4.3%	2.3%	2.7%	3.0%
NGO / NPO	4.2%	3.2%	2.3%	3.2%
Private research institution	2.6%	1.4%	1.2%	1.7%
Government institution	2.0%	2.0%	1.8%	1.9%
Private practice	1.4%	0.9%	0.7%	1.0%
International (research) organisation	0.5%	0.6%	0.2%	0.4%
Other	1.2%	0.2%	0.2%	0.5%
Total	**100.0%**	**100.0%**	**100.0%**	**100.0%**

Rank

Among the respondents employed within the ranks of higher education institutions (Table 9), 43% hold the rank of professor or associate professor, followed by 27% at the senior lecturer level and 25% at the lecturer rank or equivalent. As one would expect, the disaggregation of rank by age category reveals large differences. Only 13% of the young academic scientists hold the rank of full or associate professor. Another quarter are at the level of senior lecturer or assistant professor. The largest single proportion (48%) hold the rank of lecturer.

Table 9: **Rank (higher education sector only), by age**

	39 or younger	40–50	Older than 50	Total
Professor / Associate professor	13%	36%	75%	43%
Senior lecturer / Assistant professor	26%	35%	17%	27%
Lecturer	48%	25%	6%	25%
Researcher / Scientist	6%	3%	2%	3%
Post-doctoral fellow	6%	1%	0%	2%
Total	**100.0%**	**100.0%**	**100.0%**	**100.0%**

It is perhaps not surprising that rank, or more specifically differences in rank, surface in some of the interviews with young scientists. Senior personnel at an institution in West Africa are seen to be unapproachable.

> At home the teacher is like a god, that is to say I cannot talk with my dean. When it happened to me in conference to go talk with [a person] who is a Nobel Laureate who does not know me, but I went to talk with him. But if I see my dean, even say hello, I do not know if he will answer, I cannot pose a problem. (Male respondent)

A similar position is espoused by another interviewee from Cameroon. She feels that senior personnel look down on newly appointed staff.

> When someone has had you as a student, even when you become a teacher he continues to see you as a student and not as a teacher and then it gets frustrating sometimes because when you have a direction, he does not want to take that direction but to impose his. While you know that when you are in Cameroon you are your own boss, you can decide how to organise especially since it is his vision, but I do not accept that and when I publish he reads and he applauds. In fact, he said to himself that I was his student so I cannot have better ideas than him and it is this domination that I cannot stand. (Female respondent)

Interviewees are suspicious of the perceived power older academics hold over junior partners.

> The old generation cannot let go of the song. They do not trust the new generations and do not believe in transition. There is a lot of bureaucracy, people grabbing positions, projects, departments and research budgets. So they leave a lot of benefits to the younger generation and do not allow them to take advantage of it to flourish. This is a bit the biggest constraint that young researchers encounter at home in particular. (Male respondent from Maghreb country)

> So, they also have the power, in terms of your career, to either allow you to go to these trainings, or attend the conferences, or not, if they don't feel like you should … So far as they control the project. (32-year-old male from Ghana)

> Another factor is that elders, seniors means our teachers, who trained us, they do not encourage us. On the contrary, when you are collaborating with someone and you are not on good terms with that teacher, even writing to that person, he can even write to that person to cut the collaboration with you. And we have our teachers who want us to be their lick boots, you always have to lick them … there, and then excuses me for the term (he laughs) you have to follow them, you have to … you must always walk behind them, it bothers. And what do you do is you can weave a collaboration with the teacher, with someone from outside but if you collaborate such, such type, such teacher whose mentality you know, you have to break because, costs that cost, when he will learn that you are with so-and-so, he will do everything to spoil your collaboration. (Male respondent from West Africa)

Workload

Respondents reported that they work an average of 36 hours per week (with a median at 40 and a high standard deviation of 18.43). Table 10 displays the reported workload and task distribution according to sector of employment. The general workload varies from sector to sector, as does the percentage of working time dedicated to each task. Once the 'Other' sector category is excluded, respondents in the higher education sector reported the smallest number of working hours per week, while the ones working in private practice and international organisations reported working on average more hours per week than colleagues in other sectors.

Unsurprisingly, the percentage of working hours spent on teaching and supervision is much higher for respondents in the higher education sector than in other sectors. The percentage of time dedicated to research activities is the highest in private research institutions, followed by public research institutions, whereas the higher education sector holds the fifth place. As researchers in the medical sector are expected to do much clinical work, it is understandable that the proportion of time dedicated to service is the highest in this sector.

Table 11 displays the workload and task distribution according to rank among respondents in the higher education sector only.[8] Reported workload in general is lighter for senior lecturers and lecturers than for respondents in other positions. We clearly see that the percentage of time dedicated to teaching decreases as respondents rise in the academic hierarchy. As one could expect, post-doctoral fellows are able to dedicate the largest percentage of their working time to research activities, followed by researchers. The percentage of time devoted to research is much lower for lecturers, senior lecturers and professors when compared to post-doctoral fellows and research scientists.

If we control for age, we also find some noteworthy results. Older respondents work, on average, slightly longer hours per week than their younger counterparts, but only if the means are compared. Younger respondents spend less time on training or supervising postgraduate students, but more time on research, than their older peers do. It is quite possible that spending time on research could refer to the doctoral studies of the young academics.

8 Not all positions in our dataset are exhibited; Table 7 only shows only a selection of what we considered most relevant for workload-distribution issues.

Table 10: **Workload and percentage of working hours spent on each task, according to employment sector**

Sector		Working hours/week	% of time spent per year on:						
			Undergrad and postgrad teaching	Training/ supervision of postgrad students	Research	Admin and management	Service	Consultancy	Raising research funds/grants
Higher education	Mean	34.7	33.8	21.0	25.9	15.7	9.5	28.9	6.4
	Median	40	30	20	20	10	5	5	5
	Std. deviation	18.88	19.40	12.90	17.36	15.37	10.85	40.00	7.22
	Minimum	0	0	0	1	0	0	0	0
	Maximum	100	100	100	100	100	100	101	100
	N	3 239	3 491	3 386	3 736	3 100	2 721	2 601	2 286
Public research institution	Mean	38.8	10.9	15.7	50.1	19.0	9.5	31.9	10.7
	Median	40	5	10	50	13	5	10	10
	Std. deviation	16.15	13.28	13.22	24.16	17.82	10.01	40.88	10.27
	Minimum	0	0	0	1	0	0	0	0
	Maximum	100	90	95	100	100	80	101	89
	N	554	297	484	634	537	428	427	456
Private research institution	Mean	40.7	11.4	11.6	56.0	17.7	8.2	40.6	10.6
	Median	40	5	10	60	10	5	20	10
	Std. deviation	18.37	17.44	8.20	24.11	16.42	6.85	42.24	8.94
	Minimum	5	0	0	5	1	0	0	0
	Maximum	100	70	35	100	80	30	101	45
	N	74	34	46	83	72	48	61	58

Sector		Working hours/week	% of time spent per year on:						
			Undergrad and postgrad teaching	Training/ supervision of postgrad students	Research	Admin and management	Service	Consultancy	Raising research funds/grants
Business enterprise	Mean	39.9	7.4	9.1	25.7	24.9	12.8	44.4	6.8
	Median	40	5	5	20	15	5	38.5	5
	Std. deviation	15.98	9.11	9.96	22.03	23.44	18.59	36.19	8.46
	Minimum	4	0	0	1	0	0	0	0
	Maximum	95	50	60	100	90	80	101	50
	N	141	59	80	153	125	106	136	53
NGO / NPO	Mean	38.2	9.0	9.1	43.2	27.1	11.4	42.1	13.3
	Median	40	5	10	40	20	8	15	10
	Std. deviation	15.18	10.23	6.98	26.51	22.91	11.77	42.87	14.07
	Minimum	3	0	0	2	0	1	1	1
	Maximum	84	50	40	100	100	53	101	90
	N	145	50	79	156	134	96	117	107
Government institution	Mean	40	8	10	24	43	15	34	5
	Median	40	6.5	10	20	40	10	17.5	4
	Std. deviation	14.14	7.25	9.21	20.52	28.10	17.21	37.79	5.54
	Minimum	8	0	0	1	0	0	0	0
	Maximum	80	30	45	96	95	85	101	20
	N	80	36	48	94	80	63	66	26
Private practice	Mean	44.3	11.2	15.2	22.1	17.8	32.6	44.7	5.8
	Median	41	10	10	15	10	25	17,5	5
	Std. deviation	14.95	7.90	15.19	20.35	21.69	28.21	43.16	7.65
	Minimum	11	0	0	4	0	0	0	0
	Maximum	95	30	80	89	95	97	101	30
	N	42	38	35	47	40	43	40	25

Sector		Working hours/week	% of time spent per year on:						
			Undergrad and postgrad teaching	Training/ supervision of postgrad students	Research	Admin and management	Service	Consultancy	Raising research funds/grants
International (research) organisation	Mean	41.4	2.3	13.1	46.8	22.4	11.0	57.4	16.1
	Median	40	2	12.5	42.5	20	7.5	101	10
	Std. deviation	15.62	2.52	6.51	29.42	16.35	8.52	51.67	11.58
	Minimum	8	0	5	5	4	1	0	1
	Maximum	80	5	20	100	60	30	101	40
	N	19	3	8	22	14	16	9	15
Other	Mean	27.7	22.5	9.3	29.6	15.7	9.5	12.5	3.7
	Median	21	14	10	25	10	10	10	2,5
	Std. deviation	16.37	24.15	8.72	23.09	20.95	6.18	16.04	4.39
	Minimum	8	0	0	5	0	0	0	0
	Maximum	70	80	30	100	60	25	50	12
	N	21	17	15	24	17	20	15	14

Table 11: **Workload and percentage of working hours spent on each task, according to rank (higher education sector only)**

Rank		Working hours/week	% of time spent per year on:						
			Undergrad and postgrad teaching	Training/ supervision of postgrad students	Research	Admin and management	Service	Consultancy	Raising research funds/grants
Professor / Associate Professor	Mean	37.2	29.6	23.4	24.5	17.6	9.7	28.3	6.6
	Median	40.0	26.0	20.0	20.0	10.0	5.0	5.0	5.0
	Std. deviation	18.82	17.75	13.24	15.05	16.02	10.85	39.68	7.79
	Minimum	1	0	1	1	0	0,0	0	0
	Maximum	100	100	100	100	100	100,0	101	100
	N	1 332	1 468	1 507	1 567	1 335	1 156	1 076	997

Rank		Working hours/week	% of time spent per year on:						
			Undergrad and postgrad teaching	Training/ supervision of postgrad students	Research	Admin and management	Service	Consultancy	Raising research funds/grants
Senior lecturer / Assistant professor	Mean	33.2	37.4	20.6	23.4	13.9	8.9	26.1	5.9
	Median	40.0	35.0	20.0	20.0	10.0	5.0	5.0	5.0
	Std. deviation	18.9	17.9	12.7	14.5	13.1	9.6	38.7	6.3
	Minimum	3	2	0	1	0	0,0	0	0
	Maximum	98	90	90	100	85	80,0	101	51
	N	851	932	899	956	817	707	661	583
Lecturer	Mean	31.7	41.2	18.0	24.5	13.0	8.6	25.9	5.2
	Median	35.0	40.0	16.0	20.0	10.0	5.0	5.0	5.0
	Std. deviation	18.5	19.6	11.2	15.7	13.5	9.9	38.0	5.8
	Minimum	3	0	0	1	0	0,0	0	0
	Maximum	91	100	80	100	100	65,0	101	51
	N	772	858	730	871	698	629	614	496
Researcher / Scientist	Mean	36.9	15.7	19.2	44.7	19.2	14.0	44.5	10.2
	Median	40.0	10.0	20.0	40.0	10.0	10.0	17.0	8.0
	Std. deviation	20.1	15.2	13.8	24.9	19.1	15.1	44.5	10.7
	Minimum	4	0	1	5	1	1,0	0	0
	Maximum	84	70	70	100	100	85,0	101	75
	N	100	83	91	122	96	85	87	88
Post-doctoral fellow	Mean	38.9	12.2	16.2	58.2	11.2	9.8	52.8	10.2
	Median	40.0	10.0	10.0	55.0	10.0	8.0	20.0	8.5
	Std. deviation	15.3	13.6	13.2	22.2	7.9	9.2	47.1	7.5
	Minimum	8	0	2	10	0	0,0	0	1
	Maximum	70	55	60	100	40	50.0	101	40
	N	62	52	58	73	51	44	56	52

Table 12: **Workload and percentage of working hours spent on each task, according to age**

Age		Working hours/week	% of time spent per year on: Undergrad and postgrad teaching	Training/supervision of postgrad students	Research	Admin and management	Service	Consultancy	Raising research funds/grants
39 or younger	Mean	34.1	36.4	17.6	29.7	13.3	9.0	32.9	7.2
	Median	40.0	35.0	15.0	25.0	10.0	5.0	8.0	5.0
	Std. deviation	18.8	20.4	11.8	20.5	13.3	11.2	42.4	8.8
	Minimum	0	0	0	1	0	0.0	0	0
	Maximum	100	100	100	100	95	95.0	101	100
	N	848	910	822	968	768	676	679	611
40–50	Mean	34.8	34.8	21.1	24.3	16.0	9.6	26.7	6.1
	Median	40.0	30.0	20.0	20.0	10.0	5.0	5.0	5.0
	Std. deviation	19.2	19.1	12.3	15.5	15.6	10.4	38.6	6.6
	Minimum	3	0	0	1	0	0.0	0	0
	Maximum	100	100	90	100	100	85.0	101	70
	N	1 290	1 402	1 359	1 460	1 243	1 098	1 022	920
Older than 50	Mean	35.6	31.1	23.1	24.4	17.6	10.0	28.5	6.1
	Median	40.0	30.0	20.0	20.0	10.0	5.0	5.0	5.0
	Std. deviation	18.6	18.6	13.6	15.9	16.5	11.3	39.6	6.4
	Minimum	1	0	1	1	0	0.0	0	0
	Maximum	100	100	95	100	100	100.0	101	60
	N	1 032	1 106	1 125	1 211	1 020	885	835	703

Interviews with young scientists clearly showed that the main contributor to an excessively burdensome workload appears to be the large size of classes and too many postgraduate students requiring supervision. In addition, interviewees indicated that they are burdened with heavy administrative demands that are often not related to their core academic functions. Deans and heads of departments carry additional administrative loads.

Heavy teaching loads

A recurring complaint is related to the large workload of scientists in Africa.

> Even if the university is willing to subsidise the research of its professors, they are not ready to reduce the other workloads. The hours of lectures or the administrative charges must remain the same, that is to say the researchers must always teach the same number of courses, always be present for administrative activities. The research they do must be done in their extra time. And so for them it's a big constraint. (Male respondent from a Maghreb country)

> Well, there is the workload that I can mention because here my daily work was quite heavy, rather restrictive, and if you had to develop research, you really needed goodwill. So here we are, the mass of work. (Male respondent from West Africa)

Respondents emphasised the unsustainable number of students in their classes.

> I taught the first years and the second years and the numbers were just crazy. So, one lecture hall which could only fit 750 students, managed to squeeze in 1 000 students or you know 1 000 students just started to come in there. And the numbers were about 2 500 and that was after, you know, students like to choose their first preference and that was after the first preferences were gone. And I mean initially it started with 3 000 … making sure the venues were there, the lecturers were prepared, the tutors were prepared. So, you just run out of time and being expected to do research, was just ridiculous. (29-year-old female from South Africa)

> Teaching many undergraduate students. In Kenya, we have so many undergraduates in one class. Like currently I am teaching an undergraduate class, fourth-year groups, that is about over 800 students. I am one lecturer. And there are no tutorials, there are no marking assistants … So the one thing that is, it doesn't allow me to do any research on my own, my solicited visa is very minimal. It is very minimal because of that workload. (40-year-old male from Kenya)

The large numbers are not confined to the undergraduate level.

> The postgraduate students in reality you can't be working with 40 in a class. For some of us we would want to be able to go along with the current state of the action, techniques and carrying out of our research. (34-year-old male from Nigeria)

Relatedly, numerous interviewees felt their institutions were understaffed.

> But as far as that is concerned, we have policies, and these are in place, but we don't even, to meet the criteria, we are 25% staffed. It's not 100% as it should be ... They know the workload and they know that they are understaffed. It's understaffed, and they are not even making anything that could provide time for us to do the research. So we end up with minimum opportunities. (34-year-old male from Uganda)

> We are understaffed. We don't have enough staff for the work. So it's like we are overloaded, many students and we are looking at a better way to handle this because we cannot employ professors on the grant fund. No organisation will give you a grant to pay for the salary of the professors. That is the responsibility of the government. So there are things that they are struggling with. (40-year-old male from Tanzania)

> They would have to hire more people. Especially the young universities that have just started up, they have very few PhDs in every department, so the PhDs that are there have to do the admin as the others are probably doing their studies, trying to get PhD or masters or other degrees. So, in my department there's just two PhDs, so one is the Dean and then the I'm Chairman, so everybody's doing admin. So, the only thing they can do is more people. (33-year-old male from Kenya)

Some interviewees stressed that there is an unequal distribution of teaching loads between junior and senior members of staff.

> I know for other junior colleagues such as myself, we were just there being a little bit confused because the senior guys that would be able to do the research and reduce their teaching workload, which meant us junior people would take over the teaching workload. Without realising that we need time to do the research, but we don't have any kind of research credits to fall back on like the senior colleagues. So, we would kind of be stuck and then there was the stress of, you know, students are the priority and you have to do very good teaching and all of that stuff. (29-year-old female from South Africa)

> The argument, at least in my department, is that because the senior profs contribute so much more to the research at the department and thus inherently carry more of the research load, the junior lecturers thus need to carry more of the lecturing load. That's sort of the rationale. ... Nobody wants

> the big courses. So to some extent you sort of have to work your way up to the smaller courses, the more technical courses, the less administratively heavy courses. ... The most difficult courses to teach, are the first- and second-year courses because they're big and the students know the least amount, they're the least professional. And you would, there's an argument that the most experienced lecturers and most experienced teachers should take those courses. So then, instead we as young lecturers get those courses. You sort of very much get thrown into the deep end, you know. So now you have the biggest courses with the most admin workload, that you have to now very quickly learn about everything. (31-year-old male from South Africa)

Numerous interviewees discuss this tension between teaching and research and many researchers are forced to undertake research outside of normal working hours. Furthermore, various academics would prefer to focus the majority of their attention on either one of these two activities.

Administrative duties

Many respondents cited the time spent on teaching-related administrative duties as a major barrier to research. These include multiple meetings with moderators and external examiners.

> The undergraduate teaching comes with a lot of administrative functions ... Having meetings with your tutors, having moderation meetings with marking assignments and marking tests, setting the exam papers, setting the supplementary exam papers, communicating with the external examiner, it's sitting through the exam meetings and it's a whole lot more than just the actual teaching. (38-year-old female from South Africa)

Respondents highlighted the burden of student registration and counselling students.

> So, I'm saying that's why I said to you again, if we are not involved in registration for two weeks in January, we can spend that time attending a workshop instead of spending time at registration. (44-year-old female from South Africa)

> The other thing is there's a lot of student counselling that goes on. Whenever the students hit an academic glitch, you're the first port of call for that. (38-year-old female from South Africa)

It is expected of lecturers to be a part of various committees and boards.

> In addition to that comes a lot of administrative duties and one is always required, as an emerging researcher, to start joining committees, faculty committees and, you know, participate in, there are a lot of auxiliary things

> like organising conferences and workshops and things. A lot of these things are highly counter-productive to the research project … Although my position is supposed to be geared towards producing research, I would spend at least 60% or 70% of my time doing non-research activities … In a week, five working days, at least three and half of those days, and sometimes … four of those days would be spent doing administrative duties and I would be left with maybe one day that I could use to actually do some research. (35-year-old male from South Africa)

> Besides the extension work, there is also your institutional commitment, serving on one board, serving on the thing, coordinating PGs and all that. It takes a chunk of your time in work. So at what point do you end your work and then going to write up the research proposals and all that. (38-year-old male from Ghana)

Academics occupying positions of responsibility within their departments and faculties also report correspondingly higher levels of administrative duties.

> Admin, I'm a Chair of Department, because I was at that time, so admin means handling everything to do with the department, handling budgets, handling duty allocations and monitoring, handling student issues, handling, you know, assigning teaching duties, you know, admin … I don't know if you're familiar with university administration. (33-year-old male from Kenya)

> I do have some time to do research. I've only been restricted because I have more administrative assignments … I'm the Acting Head of my department. (35-year-old female from Nigeria)

> Being head of department then your administrative involvement is a bit high … You are called for meetings sometimes after midnight. So you have your very good plan the day before and at midnight you get this text message saying can we meet tomorrow in the boardroom at seven. (40-year-old male from Kenya)

> Currently now that completes my term for dean … I think that one was heavy in terms of administration; basically, I would say almost 40–50% of my time is generally more administrative. (36-year-old male from Zimbabwe)

Interviewees cited the time spent on administrative duties, which are not considered to be part of academic work, and therefore, they felt, should be carried out by the university administrators.

> Yes, a lot of administration and a lot of student support that is properly the job of the university administration, so just one example is being bogged down by student queries relating to things that we can do nothing about because the

> administration simply doesn't get back to them and if they do, it's a question of, like, six weeks, two months type of time-frame … [We] have to mediate with the administration, which is not really an academic function. (29-year-old male from South Africa)

> Most of my time during the day is taken up by my office-related duties which are not related to the academy. (29-year-old male from Ethiopia)

Burden of supervision

Postgraduate supervision is considered to be especially burdensome.

> And then, I think, the other barrier would probably be the postgraduate supervision because our students are at different points in their study. It's kind of managing where they are. And their topics are also really diverse … That ability to have to switch between those content areas … does impact and is a barrier to my own research particularly because it's not as clearly aligned to my own study (43-year-old female from South Africa)

A large number of students are being supervised by academics in Africa.

> Supervision, I do a lot … I do it, a lot of postgraduate students, a lot of masters. Actually right now masters that I supervise right now are about ten … And the PhDs are more than that. The PhDs, I have more than ten now. Actually, I have over 15 PhD students now, a big lot … Time for my own research is an issue because I spend most of my time reading students' theses and giving feedback. (40-year-old male from Kenya)

Supervisory-related obligations are not confined to postgraduate students. Undergraduate supervision does not provide the opportunity to undertake research.

> And of course, the workloads. The workload is still an issue. You have to teach more and also supervise many students. Sometimes the supervision of many students can be interpreted to mean maybe opportunity to conduct research. But in this case, many students are undergraduates, so they're just learning how to do research. (40-year-old male from Kenya)

Balancing the tension between teaching and research

There is clearly a tension between teaching and research-related activities for scientists in Africa. Although promotion criteria may require specified research outputs, institutions seem to be placing increasing pressure on young academics to focus on teaching-related duties. There is a clear trade-off between these activities for interviewees, who report little to no time and support to conduct research. An unequal share of the teaching burden appears to fall on newly appointed staff at various institutions.

As alluded to in many of the quotes above, the excessively large number of students in African institutions, and consequent teaching loads, have severely constrained the capacity of scientists in Africa to undertake research.

> But now we have 115 students in that lab. So it means the system put together by the Americans to contain just 30 students … now has 115 students … The issue now is, when they write exams, I have 350 to mark … In some of the courses we have two assessments; some we have one assessment. Hardly any have one, hardly. Mostly, at least, is two … If I had time I would probably do far more work than I have … Because, for me, research is my hobby. I love doing it. I have so many papers. There are so many I want to complete but I don't have the time. In fact, now they are sitting there in an envelope next to all the student papers I have to mark. And my papers are there, I want to write them, but I am not writing them. (32-year-old male from Nigeria)

> I would say about, the admin is a bit more now, both teaching and admin, if I'm honest, I think they take up 100%. Any research I do, I have to try and do it outside, you know, after, I work already maybe eight hours a day, that would be solely teaching and admin. (33-year-old male from Kenya)

A newly appointed academic describes the tension between the teaching and research requirements.

> I'm new here and I'm trying to find my feet here. So, after last year, I would say I was publishing quite well. But now, this year, with teaching and everything, I've managed only to complete I think two articles. And these are articles I actually started writing last year. There is one further article but I have not been given time to complete. I'm now even behind with it because of this period of time. So basically, I can say that this year, I didn't start any new publications because of teaching. Articles were completed in March, where actually articles were started last year but I completed it in March. And there is one that I also started last year but I haven't been able to finish. So basically, the research this year, didn't start any new publications. (42-year-old male)

Although research may be a stipulated requirement for promotion, the teaching function remains the clear priority for many institutions.

> Our university is basically a teaching university … There is no formal time allotted to doing research. Unless you improvise using your own means … There is no formal time and resources … The pressure, yes, there is pressure, but it is not very direct. Can I elaborate? … What I mean, for example, to get a promotion, okay? You need to demonstrate that you have done research and you have published it, okay? But there is no special time or resource allocated to that. So it is up to you to decide to do research and get to the next level. (37-year-old male from Uganda)

The emphasis on teaching is reflected in budget provisions.

> Because our budgets are so strict here in terms of what we can buy and in terms of assets and things, and we don't get much money. ... So teaching and learning is prioritised over research budgets. So if there is money to be spent, it usually goes to teaching and learning, which is obviously our primary business, so it's not a bad thing, but that often leaves little money over to build research infrastructure or equipment that could benefit research. (40-year-old female from South Africa)

It is therefore not surprising that many young scientists in Africa believe that research is not as prioritised as it is in developed nations.

> Yes, you are right, because as far as I know in the West or in developed nations, you know, university professors are expected to publish otherwise they may not work in universities. But here, whether ... or not, whether we do research or not, they are not, it's not mandatory to stay in the university, doing research is not a prerequisite to stay in their academic area. (36-year-old male from Ethiopia)

> I mean, the emphasis on research is not as strong as it is in the UK. More or less when a lecturer is employed here, the most important reason is for him or her to be able to teach the courses. Research is just seen as a means of maybe getting publications and getting promotion but it's not seen as an integral part of a lecturer's duties. So, not much space is really given for research, actually. (38-year-old male from Nigeria)

Institutions in Africa reportedly do not provide the opportunity to focus on research activities.

> Every country's higher institution have their own peculiar features and peculiar worries and concerns, you know. In our case I think the workload is a lot of the problem, teaching responsibility needs to be balanced, you know. I was discussing with a colleague of mine ... And she was telling me that she is moving from the University of Lancaster to Manchester or Manchester to Lancaster, I wasn't sure which way, you know. And I was wondering, I was, like, okay why are you moving, why are you changing, what's the problem, I mean, you seem to be at a pretty nice place now. And she said, look I just finished my PhD and have a lot in terms of my research commitment, so I want to do a lot of research, but my present gives me more teaching, you know, and for me I want something that would give me more research and less teaching. So, you know, she got that opportunity and then, you know, she was able to switch places and go to a university where she would teach 30% and do 70% research, which is her preference. So, I think in African countries if we have that kind of flexibility I think it will go a very, very long way in improving, you know, the overall output or capacity of different individuals and researchers and teachers as well. (34-year-old male from Nigeria)

In conclusion, it is clear both from our survey study and the interviewees that young scientists have excessive workloads. They are often assigned huge teaching loads which impact negatively on their time to conduct research. It also seems as if the relationship between teaching and research is not well managed and that conflicting signals are sent to young academics about what to prioritise.

The career challenges young scientists face

One of the main aims of the study was to ascertain what the barriers are that young scientists and scholars in Africa experience as far as their academic and scholarly careers are concerned. In terms of contextual factors, national career systems (Musselin 2002, 2010; Pezzoni et al. 2012) are particularly relevant for the careers of young scientists. In this regard, formal/legal aspects, such as short-term contracts, the organisation and execution of appointment processes for professorships, departmental strategies, and national academic labour markets, are identified as relevant factors that affect career development. Musselin (2002) points out that the definition of what constitutes a 'quality' or a 'good' candidate for a job in higher education – in terms of scientific activities, personality/collegiality of the candidate, and teaching abilities – can vary greatly from one department to another and evolve over time (Musselin 2010). Hence, the evolution of department recruitment rationales, the continuous assessment of performance and the development of relevant promotion criteria are also important factors to consider and must therefore be taken into account in higher education research (Barrett & Barrett 2008; Farnham 2009; McInnis 2010; Leisyte & Dee 2012; Leisyte & Westerheijden 2014).

In recent years, non-structural, subjective factors such as normative orientations, career expectations and alternative career options for PhD holders in other employment sectors have received increased research attention. Roach and Sauermann (2010) highlighted the importance of individual preferences of PhD students in science and engineering, and outcome expectations of individual career choices after graduating, as resulting in various career paths. Jaksztat et al. (2010) identified three distinct 'mindsets' that lead to different careers. Academic researchers focus on academic research, striving for employment within higher education or research institutes, but have a lesser interest in private enterprises; exit-oriented researchers have the broadest range of vocational orientation, i.e. a lesser interest in research, but striving for a combination of research and innovation in the private sector; and application-oriented researchers have a strong focus on research, and a strong focus on the private sector, i.e. if interested in academic research, they prefer research organisations to universities. Role-models may also have strong influence on these mindsets, e.g. co-workers in a department who had previously worked in the private sector, and co-authors who had become academic entrepreneurs, may increase the likelihood of academic scientists becoming academic entrepreneurs themselves (Stuart & Ding 2006).

In order to gauge what the main challenges are that young scientists face, we asked respondents to indicate to what extent (not at all; to some extent; or to a large extent) 10 predetermined factors may have impacted negatively on their careers as (academic) scientists thus far. They were also asked to add any other factors to this list.

Table 13: **Respondents' perceptions of the impact of 10 factors on their careers (challenges ranked)**

Challenges to their careers ...	Overall rank	Rank by age		
		39 and younger	40 to 50	Older than 50
Lack of research funding	1	1	1	1
Lack of funding for research equipment	2	2	2	2
Balancing work and family demands	3	6	3	3
Lack of mentoring and support	4	4	3	4
Lack of mobility opportunities	5	5	5	5
Lack of training opportunities to develop professional skills	6	3	4	6
Lack of access to library and/or information sources	7	7	6	7
Limitation of academic freedom	8	9	7	8
Job insecurity	9	8	8	10
Political instability or war	10	10	9	9

The results in Table 13 show that general lack of research funding and funding for equipment were identified by all respondents as posing the biggest challenges. Given our discussion in Part One of this book about the lack of especially national sources of funding, this is not a surprising result. The third largest challenge (balancing work and family demands) speaks to time demands. This was identified by all respondents – except for the youngest cohort (39-years and younger) – as the third largest challenge. Challenges related to human capacity building and professional development (lack of mentoring and support; lack of mobility and lack of training opportunities) were subsequently listed as the next largest challenges by most of our respondents and, importantly, especially by the young scientists. The fact that lack of library and information resources was rated relatively low, may be an indication that many scientists and scholars now have access to their information sources through the internet rather than relying solely on local libraries and collections. Political and social factors (political instability and lack of academic freedom) received the lowest rating. Interestingly, 'job insecurity' was also not listed as a major challenge – probably because most of our respondents are already in permanent academic or research positions.

Many of the other challenges respondents themselves identified refer to the difficulties of managing multiple tasks. Administrative duties, teaching load and clinical work limit the time available for research (in this regard, sometimes respondents also refer to additional employment or the necessity to have consultancy contracts to increase income). Another important category concerns the problems of bureaucracy and administration, institutional politics, and nepotism. Respondents also point out various types of discrimination (affirmative action or racism; gender and age discrimination; xenophobia; religion; or ethnicity). Funding was again mentioned in response to the open-ended question, along with the lack of resources for research. Respondents also highlight a lack of human resources, either a lack of good technicians and students, or a lack of peers and researchers. In this regard, they also stress their isolation and the lack of collaboration.

In the remaining chapters of this section we discuss the main challenges as rated by our respondents in more detail. In Chapter 5 we look more closely at the issue of funding and also why the lack of funding was identified by the majority of the sample as having had the biggest negative impact on their careers. This is followed by Chapter 6, which focuses more specifically on issues related to mentoring and lack of support for professional development, whereas Chapter 7 is devoted to a discussion of the challenges related to mobility. The final chapters of this section then focus on how various biographical and career-related aspects impact on the research performance of young scientists, and specifically on their research output (Chapter 8), as well as their networks and research collaborations (Chapter 9).

CHAPTER 5

Lack of funding

Catherine Beaudry, Johann Mouton and Heidi Prozesky

Introduction

Investment in science in Africa remains low by international standards. As shown in Part One of the book, government investment in R&D as proportion of gross domestic product across Africa averages between 0.2% and 0.3%. This is despite the fact that in 2005 most ministers of science and technology in Africa committed themselves to a target of 1.5%. The low investments in science and research by African governments also mean that many countries are heavily dependent on foreign funding for research. Investments by international governments, donors and development agencies in research have increased significantly over the past two decades. However, these investments are skewed towards health- and agriculture-related areas, and often reflect the priorities of the funders.

The link between public funding of research and scientific production is well established in the literature. Payne and Siow (2003) as well as Blume-Kohout et al. (2009), for instance, show that the public funding of university research has a positive effect on scientific production at universities. Furthermore, attracting public funding for specific projects can be perceived as a signal of quality, not only of the funded researchers, but also of their university. Adams et al. (2005) show that top universities and departments that have been awarded public grants have larger teams (with an increased scientific division of labour at the international level) and attract more government funding. Pavitt (2000, 2001) also highlights the importance of public support for scientific infrastructure development in the US.

Some studies have shown that better-funded scientists are more frequently cited and more productive than less-funded scientists (Beaudry & Allaoui 2012). The granting of research money further acts as a signal that attracts additional funding in subsequent years. Arora et al. (1998) show that the publication track-record of researchers has an influence on future grants and, consequently, on future publication levels as well. Zucker et al. (2007) show the major positive impact that research financing has on the number of scientific articles published. Jacob and Lefgren (2007) find that specific grants add one additional publication within the five years subsequent to the attribution of the grant.

In addition, collaboration can become a powerful lever to raise funds (Daniel et al. 2003), and consequently, scientific collaboration and research funding are intrinsically intertwined. Multi-project research centers encourage researchers and their universities to collaborate more efficiently, thereby leading to a more efficient use of the available diversity of resources of a physical, human and/or financial nature (Zucker et al. 2007).

As discussed in Chapter 4, the survey respondents listed a lack of research funding and a lack of funding for research equipment as their two largest challenges. In both cases more than 50% of the respondents indicated that they had experienced these to a large extent. A number of other questions related to research funding were put to respondents. The first was aimed at establishing whether they had received any funding over the preceding three years.[9] This was followed by questions about the average amount of funding received and the sources of funding. We now turn to a more in-depth analyses of our findings on research funding.

Funding received

In response to the question whether our respondents had received any funding for research over the preceding three years, slightly more than half (55%) responded in the affirmative. Somewhat surprisingly, the disaggregation by scientific field did not reveal huge differences (Figure 33).

Figure 33: **Receipt of funding (Yes), by field**

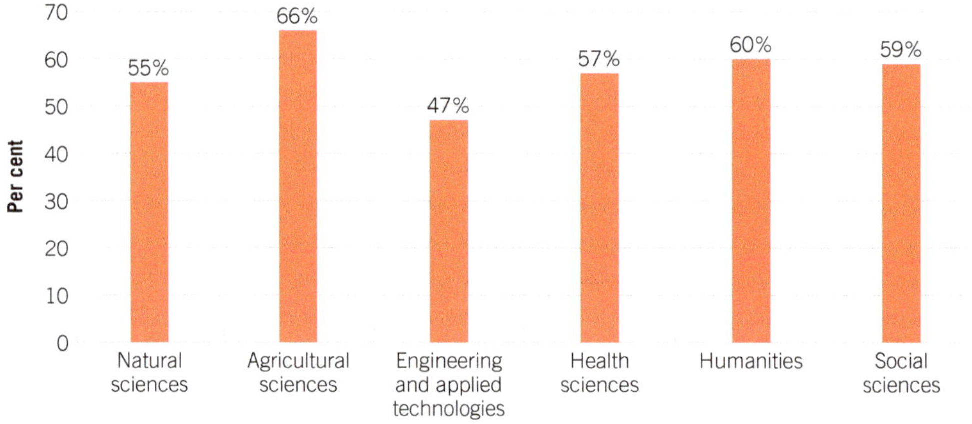

We ran a three-way analysis of variance with age interval, gender and scientific field as predictor variables (Figure 34). Age and scientific field were found to relate significantly to whether a respondent had received funding. In general, older (especially those over the age of 50) respondents across all fields reported having had received funding. As far as field differences are concerned, respondents in the health and agricultural sciences, and perhaps unexpectedly, also the humanities, were more likely to report that they had received funding.

9 We are aware that funding information obtained by the survey is likely to be biased by a misunderstanding of the question. We asked the amount of funding individuals personally received over the last three years (i.e. preceding the survey), but we realised that many of them understood the question to be not solely about themselves. For instance, some of them provided us with the total amount of funding received by their team or their laboratory.

Figure 34: **Reported receipt of funding, by age, field and gender**

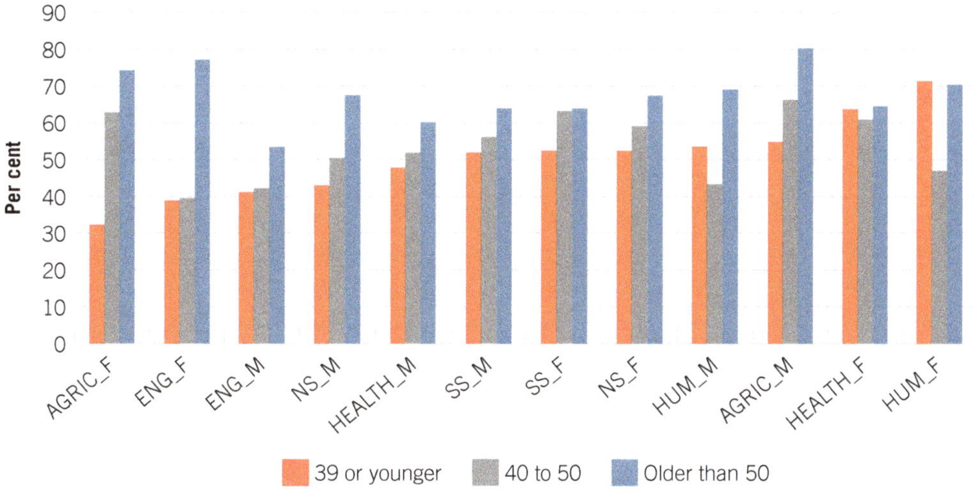

Amount of funding received

The interval-level variable 'amount of funding received' was converted to a scale variable, by calculating the mid-point value of each interval, and assigning that mid-point to the respondents in each interval. In the case of the highest category (more than US$ 1 000 000), three approaches were followed, ranging from most to least conservative estimations of the mid-point value: (1) define the mid-point as US$ 1 000 000; (2) presume the same interval of US$ 500 000 that we find in the second-highest category, which leads us to calculate the mid-point for the highest category as US$ 1 250 000; and (3) define the mid-point somewhat higher, at US$ 1 500 000. All three approaches produce the same median of US$ 5 000, but the means and standard deviations differ. In order to minimise any bias in our reporting, we followed the first (most conservative) approach in estimating the mid-point values of funding received.

Table 14: **Reported amount of funding received by field**

		Option 1	Option 2	Option 3
N	Valid	4 825	4 825	4 825
	Missing	212	212	212
Mean		73 154.92	79 787.05	86 419.17
Median		5 000.00	5 000.00	5 000.00
Std. deviation		197 058.87	229 658.51	264 353.22

It is also important to point out that the distribution of values is very heavily skewed (positively). In other word, a relatively small number of respondents listed very high funding amounts, whereas the majority reported amounts of less than US$ 100 000. Figure 35 shows the distribution of those respondents (approximately 55% of them) who indicated that they had received funding in the preceding three years. The surprising result is not the fact that the distribution is positively skewed; one would always expect that the majority of respondents would indicate having received more moderate amounts (approximately 44% had received less than US$ 100 000).

Figure 35: **Distribution across categories of reported amount of funding received (US$)**

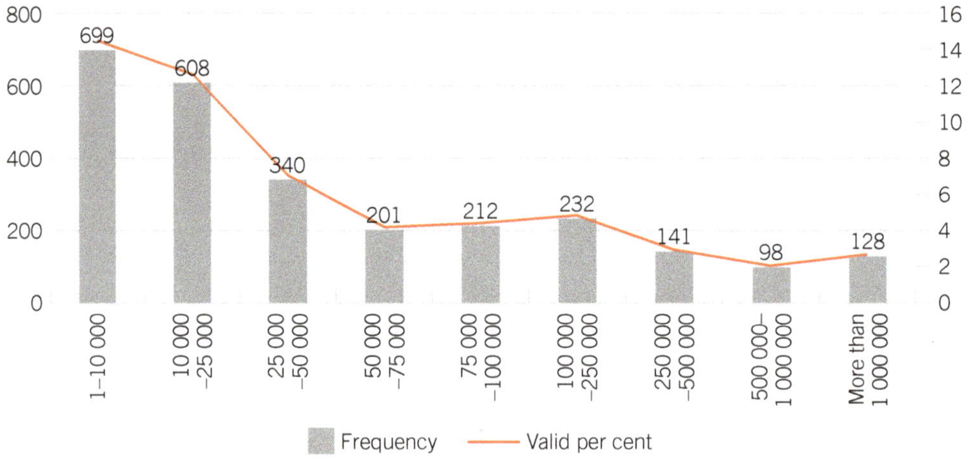

What is quite striking, though, is the fact that 128 respondents indicated that they had received more than US$ 1million. Close inspection of the profile of these 128 respondents revealed the following: the gender distribution for this group is commensurate with that of the total sample (70% male / 30% female); they are mostly over the age of 40 (but it is noteworthy that 19 [15%] of them are young scientists); the largest proportion are resident in South Africa (40%), followed by scientists from Kenya (14%), Uganda (6%) and Zambia (6%). Perhaps not surprisingly, the largest single group are from the health sciences (34%), followed by scientists from the natural sciences (22%), agricultural sciences (20%) and the social sciences (14%).

Disaggregation by age (Table 4) revealed the expected result, with older respondents on average reporting much higher amounts of funding received.

Table 15: **Reported amount of funding received, by age**

Age	Mean	Median	N
39 or younger	47 286.59	0	1 394
40–50	79 800.65	5 000	1 856
Older than 50	91 847.53	17 500	1 456
Total	**73 896.62**	**5 000**	**4 706**

A breakdown by field (Table 16) again revealed a result that would conform to our expectations, with respondents in the agricultural and health sciences reporting the highest average amounts (by some margin). The relatively low amounts reported for the other fields conform to the results presented in Chapter 3 of this book.

Table 16: **Reported amount of funding received, by scientific field**

Field	Mean	N	Std. deviation
Agricultural sciences	105 367.91	564	234 814.96
Health sciences	95 778.35	1 076	234 436.39
Social sciences	66 761.24	934	178 847.62
Engineering and applied technologies	59 770.68	556	174 028.46
Natural sciences	58 162.03	1 401	171 967.67
Humanities	44 740.74	270	153 849.99

A three-way ANOVA showed that age of respondent and field were the strongest predictors of differences in amounts reported. Gender by itself, as well as in interaction with age and field do not seem to be highly correlated with reported funding amounts.

Figure 36: **Reported amount of funding received, by age, field and gender**

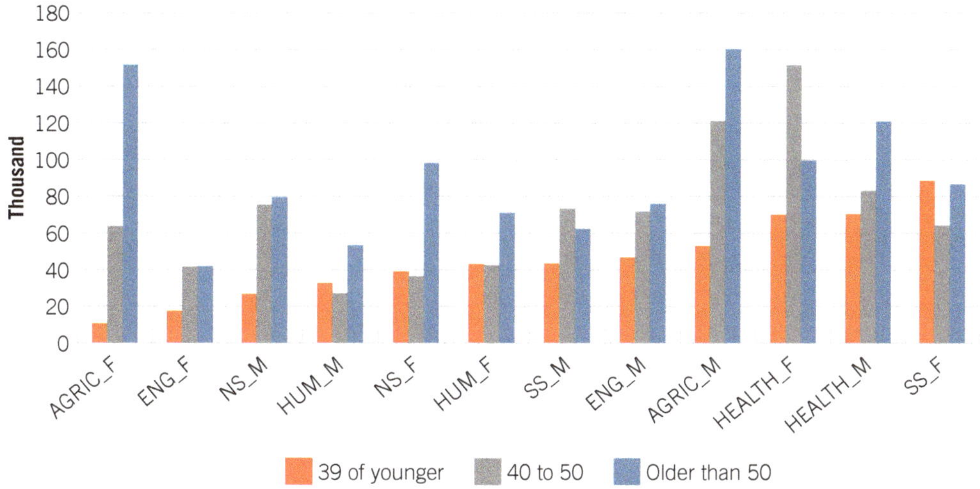

Sources of funding

In Part One of the book we discussed the fact that African research continues to rely heavily on foreign sources for funding. We asked survey respondents to indicate what proportions of their funding are sourced from international and national sources respectively. If we focus on the young scientists only, the results are interesting (Figure 37). They show that as the proportions of funding increase, the amounts respondents received increase and more so from international sources. Stated differently: Those respondents who indicated that more than 50% of their funding come from international sources also recorded significantly higher average amounts of funding received.

Figure 37: **Proportions of funding sourced from international and national sources (39 or younger only) (US$)**

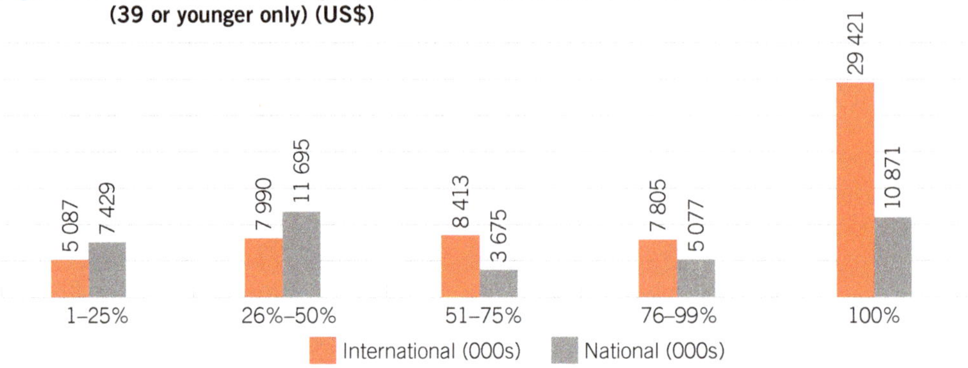

In the remainder of this section, we present more information on the reported funding received from national and international sources disaggregated by field (Tables 17 and 18) as well as by age, field and gender (Figure 38 and 39).

Table 17: **Reported percentage of funding from national sources, by field**

Field	N	Median
Engineering and applied technologies	258	100
Humanities	149	100
Natural sciences	730	90
Social sciences	497	60
Agricultural sciences	335	50
Health sciences	562	50
Total	**2 531**	**80**

Figure 38: **Reported percentage of funding from national sources, by age, field and gender**

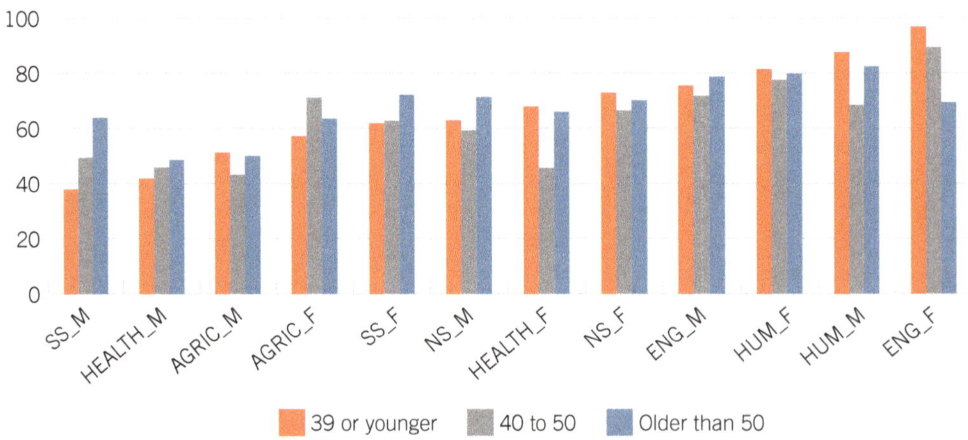

Table 18: **Reported percentage of funding from international sources, by field**

Field	N	Median
Health sciences	513	80
Agricultural sciences	326	70
Social sciences	435	70
Natural sciences	588	50
Humanities	97	30
Engineering and applied technologies	192	10
Total	**2 151**	**60**

Figure 39: **Reported percentage of funding from foreign sources, by age, field and gender**

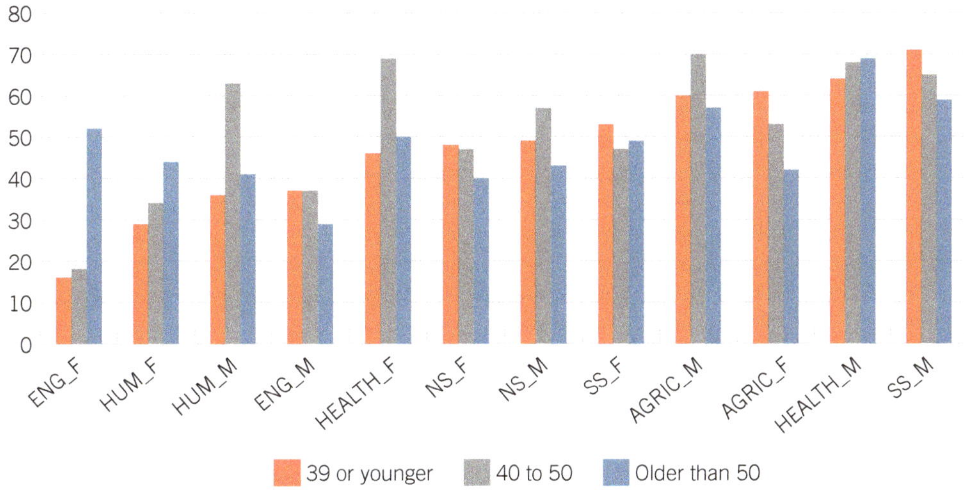

Major funding organisations

We also asked respondents to name the three main funding organisations from which they had received grants. Some of the responses were impossible to identify, because respondents provided only general information, such as 'funding agency', 'science council', 'business enterprise', 'university' or 'government', without specifying the source in more detail. Nevertheless, we report on the funding organisations most frequently mentioned by the respondents.

The South African National Research Foundation is by far the most cited funding organisation from which the scientists had received grants, followed by the European Union. The sources of funding are, however, very heterogeneous: we find public, private and non-profit organisations, as well as local, national, international, and overseas institutions. These results correlate with the analysis of funding acknowledgements as presented in Chapter 3.

Table 19: **Most frequently mentioned funding organisations**

First funding organisation listed		Second funding organisation listed		Third funding organisation listed	
National Research Foundation – South Africa	773	National Research Foundation – South Africa	158	National Research Foundation – South Africa	57
European Union	89	European Union	54	European Union	32
National Institutes of Health – USA	86	Government – South Africa	50	USAID	19
Ministère de l'Enseignement Supérieur et de la Recherche Scientifique – Algérie	83	The Bill and Melinda Gates Foundation	38	South African Medical Research Council	17
Government – South Africa	74	USAID	34	World Bank	16
Government – Nigeria	64	South African Medical Research Council	33	The Bill and Melinda Gates Foundation	13
Ministère de l'enseignement supérieur et la recherche scientifique – Tunisia	59	Department of Science and Technology – South Africa	25	IDRC – Canada	11
Tertiary Education Trust Fund – Nigeria	52	DFID – UK	23	National Institutes of Health – USA	11
USAID	45	National Institutes of Health – USA	22	DFID – UK	10
The Bill and Melinda Gates Foundation	43	IDRC – Canada	21	WHO	10
IDRC – Canada	34	Ministère de l'enseignement supérieur et la recherche scientifique – Tunisia	19	Wellcome Trust	9
DFID – UK	29	World Bank	19	IRD – France	8
Wellcome Trust	28	Wellcome Trust	18	Agence Universitaire de la Francophonie	7
Direction Générale de la Recherche Scientifique et du Développement Technologique – Algeria	26	Center for Disease Control – USA	16	Medical Research Council – UK	7
International Foundation for Science	24	Direction Générale de la Recherche Scientifique et du Développement Technologique – Algeria	16	National Geographic Society	7
Water Research Commission	24	Water Research Commission	16	BMZ – Germany	6
World Bank	24	University of Cape Town – South Africa	14	DST – RSA	6
Agence Universitaire de la Francophonie	20	Government – Nigeria	13	International Fund for Agriculture Development	6
South African Medical Research Council	19	WHO	12	Ministère de l'Enseignement Supérieur et de la Recherche Scientifique – Algérie	6
Center for Disease Control – USA	17	Agence nationale de la recherche – France	11	Ministère de l'enseignement supérieur et la recherche scientifique – Tunisia	6
CNRST – Morocco	17	IRD – France	11		
University of Cape Town – South Africa	17	Stellenbosch University – South Africa	11		
Agence nationale de la recherche – France	16				

Barriers to securing funding and the consequences

During the personal interviews, we probed participants on a number of issues related to funding. A salient issue that emerged is the difficulty that they have in obtaining funding for research purposes. As the responses show (especially from the respondents from Nigeria) the nature and extent of the challenges involved in securing funding vary from country to country.

> Extremely difficult, to be honest. Extremely difficult, in fact most times, for instance, between 2016 and now I've only gotten, it's more, like, saying maybe consultant fees, but in terms of funding research … it's very, very difficult, when you apply it's so, I try hard, I write proposal, I try to follow up the proposal, put all papers and all of that, I try to collaborate here and there, but … And it's quite difficult and very difficult to keep up to actually write something that is really good. Sometimes you get the calls very late, you try to meet up with the deadline and, I don't know, it's just not, sometimes it's very difficult, internally and international. (34-year-old male from Nigeria)

> As another barrier, there were some difficulties in funding research. Some financing difficulties. For your information, I took the initiative to do some work on the normal level of TD4 lymphocytes in non-HIV-infected […] people. You know at the time it was very important. It is from the bar of 500, below or above that we know if someone was immunodeficient. This figure was a bit universal. It had to be adapted to each population. And so I took the initiative to do this without any support. It was not easy but after that I received a little support from the project I was working on and it helped me. That was the second obstacle, the financial constraints. (Male from West Africa)

> You have in South Africa, where you have lots of funding bodies in Africa. In Nigeria here, honestly, it is very, very tough to have such opportunities. Yes, you can have the opportunity to … go and do masters or a PhD, but more often than that it ends at that level. (40-year-old from Nigeria)

In many instances, equipment required for research cannot be obtained.

> The biggest challenge for my career is … funds to establish research laboratories … it's very paramount. Getting funding that will help you … I mean, the machines are obsolete. You need to beef it up, okay? That is another challenge. So it means that if you are running some samples, you are running some tests, either you take it outside the country or you take it into Europe or whatever you do will not be too relevant. (36-year-old male from Ghana)

> The lack of resources, the lack of equipment, and even in the context of research there is no funding, there is no laboratory there is nothing. (Male from West Africa)

More specifically, participants identified the lack of national funding as a concern. Most funding is still received from international funders.

> Within our institution, they've tried to establish some fund for research, but it is very small and has the same issues that happen back and forth, a lot of people chasing the few resources. The maximum maybe they keep is about one to three people will succeed in the application process. You have a million people chasing something that's very hard to get, and it doesn't really make sense … We have very little government-funded research within the country. (38-year-old male academic from Ghana)

> It was an entirely other different area of my research and I had facilities, all that are valid for my research was there. The research was conducted at Jawaharlal Nehru University, New Delhi, India. So I had all the facilities that I needed for research that I conducted. Unlike here I know such research wouldn't have really been feasible here in Nigeria, yes. (39-year-old female from Nigeria)

> Most of the funding has come from maybe the UK government or some foundations. It depends on where the funding comes from, but definitely not from Uganda. (34-year-old male from Uganda)

Interviewees lament that research in Africa has become dependent upon international and donor funding.

> Some of us have over 80% of our funding by donor funding. Yet I am in a country where 85% of the economy depends on agriculture and they have over 70% of the population do agriculture. So you would expect … lots of money. (40-year-old male from Uganda)

> I am just wondering how long shall we depend on external funding, because that's not sustainable. So if the local capacity can be built and so internally if we can outsource the resources for research that could be sustainable. Meanwhile the country is struggling to fix many things – infrastructure, roads, whatever – so there is a lot of attention and focus to different diverse demands from the government, so we cannot depend on the internal support. (40-year-old male from Tanzania)

This raises the question: What are the reasons why funding is not secured? A number of interviewees attributed their inability to successfully acquire funding to inadequate qualifications.

> Another challenge is funding because currently, as head of researchers, it is difficult to secure funding because your back history is not that sanctioned to convince funders. Because, as a young researcher, your CV isn't developed

> that much. So, in that case, your funding possibilities will be very limited … Most of the research funders, as you may know, are looking at qualifications. (35-year-old male from Ethiopia)

> Yes, the funding especially because I haven't … I'm doing my PhD so when you get funding calls, then you see the funding call and it's precisely in your area and it's got a perfect research proposal that can help you start and finish. When you get to the eligibility section and oh no … should have received their PhD and have done research for at least five years or something like that and then your heart just sinks. Or if you're PhD student in that eligibility section they would say no, your supervisor needs to apply for you and then you know that your supervisor is so busy with other students and everything, so yes it makes it difficult as well. (32-year-old female from South Africa)

Other interviewees attributed their inability to successfully acquire funding to their lack of experience. Frustratingly, experience cannot be acquired without funding to undertake research.

> It's a big challenge because even the funders always require that you have some experience and then they say to be like a principal investigator, you must have done your PhD, like, five years ago or something. So you can't really get a project of your own at the moment, like, in my situation. (38-year-old male from Uganda)

> So, the research scored very high marks in terms of its scientific merit but the comments were 'inexperienced investigator'. And they were right, I am inexperienced, I haven't had any grants before, which I've done to completion. So, that's where I learn that it's even about producing a good proposal but if you don't have these other resources … (40-year-old male from South Africa)

An inability to secure an international partnership also led to ineligibility for funding.

> As a country, we have limited resources like for funding for research, but of course, we also try international-wise. Now international-wise, you have to have at least some collaborators from other countries … So the proposal might be good, but because we couldn't have a partner from that country, it couldn't go through. So we couldn't get any. (39-year-old male from Tanzania)

In South Africa an interviewee was allegedly deemed ineligible due to non-citizenship.

> I'm not a South African citizen. Sometimes that also is a barrier on my behalf, as they may not consider my application as compared to an application by a South African. (36-year-old female from South Africa)

Bureaucratic hurdles and inadequate research committees are also blamed for funds not being made available.

> Mostly administrative because in order to obtain funding, we have to go through several research committees, which generally do not work with a transparent mechanism but a rather bureaucratic mechanism, the old way and therefore sometimes these commissions do not work. [They] do not understand the scope of the project because it is formed of people who do not have the same specialty and suddenly it is among the obstacles to obtaining funding. (Male respondent from a Maghreb country)

Interviewees listed a number of negative consequences that result from a lack of funding:

- researchers are less productive;
- researchers constantly have to devote their time searching for funding; and
- some research fields simply do not obtain funding.

An interviewee describes how lack of access to research equipment leads to lower productivity.

> With me, of course we're in a fortunate position that we have good equipment, it's managed under this multiuser umbrella and of course it means that you need to budget for instrument utilisation per hour. So, overseas in many labs it's different, you have that instrument 24/7 for you, for your group. So, you could produce 24/7 data … This is a global comparison to really good labs. And … in South Africa … these are of course the constraints that we have to work with, that we have to say you know I can only budget 20 hours for you on this instrument, more if not in. Hopefully we get the data that we you know envision. (39-year-old male from South Africa)

Another consequence of the lack of resources and equipment is that young scientists are not able to obtain required skills and experience.

> Because when I went to Malaysia, I was able to view some state-of-the-art equipment which I had only read about in journals. When I went to Malaysia I was able to use them for part of my research. And that added some great discovery to my work. That's why it has been accepted by reputable and journals of high regard. That added so much to part of my work … And most of the students who actually taught me how to do things, some of them were only master's students and others were undergraduate students. They already had exposure on how to use them. On the proposal I wrote for my country, some of the students had to advise me … There are other equipment now that you can use to get positive results and better results in whatever we are supposed to do. So it actually gives them more exposure and it gives them a wider experimental agility. (39-year-old female from Nigeria)

> That's, you know, like I'm specifically training or, you know, honing my skills to be actually master of hematology. Now hematology, for example, have a lot of procedures, learned theorems. And to have, to practice … So you'll find that basically that equipment or the instruments to perform that particular procedure is lacking. (35-year-old male from Kenya)

The constant search for funding means that young scientists have less time to focus on their core functions, and this is adding stress.

> Well, it's a lot of hard work. I mean, as I've shown in that application, I've had many applications that's been rejected and I haven't had an NRF application that's been successful. But I've managed to get enough funding from especially mining companies in my field of research. And it, I mean, it's hard work, so the, the time that's left for research, a lot of it is spent just canvassing, trying to find money, trying to make sure that you meet the requirements that go along with that funding. (32-year-old female from South Africa)

> It filters down, up to departmental level, because we now have to think about how we can make more money, because the university needs to be financially sustainable. Faculty needs to be financially sustainable. So these messages of financial sustainability do filter through and they reorientate the thinking of people so that they now become aligned with thinking income-related. And then you start to carry the worries of your department and finances as well, because all of a sudden … You know, you're supposed to be an academic, you're supposed to be focusing on our students and doing your research, but you take on so much more than that, because you must now produce … you must now make money; you must do short courses so you can make some more money, the university can make some more money, you must get research grants so you can make some more money, you must … you know. You start to feel like these are the messages that come through, these are the kinds of things you must focus on. (38-year-old female from South Africa)

A number of fields are purportedly neglected areas for research funding. These include:

> *Computer science*
> Yes, funding is a challenge for me where I am, so most the funding I'm doing it myself, the initial seed funding I'm doing it myself, because my area of specialisation is in computer science and here that I am there is not so much funding to that effect. (31-year-old male from Ghana)

> *Engineering*
> So it's mostly lab equipment for the practical work and from skill experience rather than doing research. So the lab equipment for research is still lacking, especially in the engineering disciplines other than medicine. (35-year-old male from Uganda)

> *Health – neglected diseases*
> I'm looking more into neglected diseases, and these diseases are neglected primarily because, one, they don't get a lot of funding, but also there's no political incentive to look into them, or maybe they don't have that important impact in the health sector, so it is quite challenging. (35-year-old male from Tanzania)
>
> *Humanities*
> In humanities, I mean, philosophy, the humanities, humanities have limited opportunities for things like that; you have more opportunities in sciences. There's always this big gap between the opportunities available for science students and what is available for humanities or art students. Yes, more opportunities will be nice, I think there are quite a few opportunities but it's really not enough, not enough for the humanities considering the number of post grads in the universities in African countries, you know, and we really could do with such an exchange. This morning I was going online, I went on and checking for some grants to attend a conference, you know, you could check about 10 searches and eight is for science and science researchers and two is for researchers in humanities, so you have a lower opportunity of getting the funds you are looking for, the grants you are looking for, maybe for attending a conference or something. (32-year-old male from Nigeria)
>
> I have a PhD in women's and gender studies which falls under the humanities, and we all know that the humanities, it's still a side-lined, marginalised area of research globally. It would be worse in places like South Africa where we are given less money to go around for research, internal money as well as external money, in terms of getting money from other countries who have more. (40-year-old female from South Africa)
>
> *Psychiatry*
> You have to source your fund yourself. Otherwise, you apply for grants outside. Unfortunately, most of the grants here are usually HIV, malaria, polio, all these communicable diseases. I'm a psychiatrist and I'm specialising in forensic psychiatry. It's not a field that is strong here in Nigeria specifically. People don't even get what you're talking about sometimes. (40-year-old female from Nigeria)

As a result, researchers in Africa are forced to collaborate with partners in the Global North in order to secure funding for their research.

> Generally, I think it's quite challenging. We've got a couple of global funding schemes and research programmes that look mostly into pandemics and diseases like HIV and malaria. ... Within those schemes ... it is very highly competitive and not very likely that an African applying for funding would acquire that funding strictly within an African research system. Most times it

> has to be in collaboration with a western partner or some other universities that are more advanced or have better reputations of doing funded research. (35-year-old male from Tanzania)

Young scientists are often forced to pay out of own their pocket to publish, to conduct research, for further studies, to attend conferences and for internet data.

> For us here it's either you publish or you perish, so most of us here we go ahead and use our salary to do this research, so we are paying. Because at the end of seven years you have to present something, even though you don't get anything from the national system. (40-year-old female from Nigeria)

> Most times we're left to self-sponsor ourselves through universities and all that. And scholarships opportunities are very competitive, and you try applying for one, and it's not working out. So you just have to find a way to sort yourself out. So, if the opportunities, I try to apply, and see how I can make the best of the opportunities, but none have come thus far ... (30-year-old male from Nigeria)

> Well, yes, you attend conferences and seminars. Conferences in Nigeria, it's a bit busier because it's not that expensive. So you don't need to fly and things like that. So you pay for yourself really, but it's not that expensive. (38-year-old male from Nigeria)

> I have to buy more data in order to communicate. I have data, but I don't want to use it. The thing is, if I want to browse now, I have to buy the data. Otherwise the internet, it is not reliable ... To get information on the internet, I have to spend my own money ... For me to even go online, I have to buy the data with my personal money. (32-year-old male from Nigeria)

> I have time for research, I have time for research but the problem is funding, sometimes people who do research they must fund from their salary. The amount you have you fund that research, so you fund your research. Especially publication you must publish with your money and that's very difficult. (35-year-old male from Nigeria)

Recommendations: Additional funding required

Interviewees made a number of recommendations for increased funding. Funding that is specifically assigned to emerging researchers was highlighted.

> To create more funding opportunities and have them accessible to emerging researchers from all races and from all genders, and ages; support for that, that maybe have some, some funding available that's based on merit and yes, to support, yes, to support more emerging researchers. But with that being

> said, I know funding is limited, but I think, for me, the funding has been my major obstacle. The government can't really do anything more about my time, so I think that's more of a university thing and personal management, but I think, yes, funding from government. (28-year-old female academic from South Africa)

Supplementary funding is also necessary to obtain additional staff and thereby lower student-to-staff ratios.

> There needs to be more funding. I would have more time for my research if there was more funding to appoint more academics. You know, when you have a student-to-staff ratio, and this is something that is unique I think to the institution I work at. But on some modules if you have a student-to-staff ratio of one staff member per every 2 000 students, then the kind of exponential increase in all levels of your teaching work is something that does stop you from getting research done. And just developing as a scholar, going to conferences, having time to do the invisible work of reviews and so on. So I think if there was more funding to appoint more academics so that we have a larger cohort of teaching staff so that the burden isn't so heavy on the few that do teach. (30-year-male academic from South Africa)

Young scientists express a desire to work overseas, as there are better facilities available and more funds available to do research.

> There are greater chances of opportunities outside, and better funding, and career opportunities as well after the programme, or after certain trainings ... Aside from the funding ... our system here, seems limited with [available] instrumentations, and advances in certain areas ... There are kinds of work you would want to do here, but because there are no facilities for you to do such work, it becomes a problem. Whereas outside there are greater chances of going through some kind of research. And you have the funding as well. They have funding, equipment, and they are the major problems here. So considering those, if I see opportunity outside, which I do apply for one ... (30-year-old male from Nigeria)

> Because also in our places, although we don't have research funds, but also research equipment, like the big machines. I'm a chemist. ... all this equipment are not here in Tanzania. So by going abroad, you have this but you also have access to more research journals and education. So you'd be able to update your information [by going] abroad. (39-year-old male from Tanzania)

> I had a very, very good post-doc fellowship in the Netherlands. I was receiving a really excellent salary and a very good research budget and excellent conditions of employment, so I only needed to do research. And I went from a

> situation like that to a situation where I had to do, you know, 50% teaching at a salary which was less than what I was getting as a post-doc and not as much, and not any research funding, so I don't even have a research fund or a grant or anything. So, I think, and I've seen that many of my, many of the peers that I have who are really good researchers, they would just go to another institution because they definitely, you know, aren't being given a competitive package at a university or an institution, so that's why I would say that if you want to do excellent blue-sky research you have to make, you have to attract competitive people and you have to make, and you have to put out a competitive package for people. So I'll tell you something that's a little confidential, I am actually applying to different institutions, to different international entities and so on, for jobs because I simply cannot build equity given the kind of package that I have right now … (35-year-old male from South Africa)

Interviewees provided a number of recommendations to acquire funds and improve their respective funding systems. In this regard, training on fundraising was emphasised.

> I think so because sometimes you think know how, you know how to read the minds of the committees who sits to consider your applications for fund raising, but you're just, not so good. For the very first time I did apply for some funds, in a trust foundation, somewhere in the UK, to help me pursue some events on albinism in West Africa, but I recently got a reply, it wasn't successful. I feel, yes it was a good project but I just feel maybe I didn't have enough training on how to approach it. So sometimes you have to go and learn yourself and do a research, you know, what to do and what not to do, but it is always nice if your institution could regularly organise programmes to train academic staff on how to go about such fundraising (32-year-old male from Nigeria)

> If we had these short courses also, they could equip us, update information with regards to the particular research, and also give us some insights on how to write a good proposal … I think it would be more important and it could help us to get that kind of research funds, yes. (39-year-old male from Tanzania)

The reasons for funding rejections should also be made available.

> I take a lot of time to apply to the NRF, because the application process is not simple, most times. It is supported by the institution, but then often it comes back as the review has been rejected. I would like to know why it was rejected and how we … how I can improve so the next time around it can be successful … perhaps just guiding us, because I don't think that it can be rejected all the time, maybe once or so, but I think that if you take the time to apply, I know that the NRF does have funds available. … When we do apply for it, it's turned down. I don't know what the criteria is when they're reviewing or what the reason is for the rejection. (41-year-old female from South Africa)

The potential for shared subscription to journals (between universities) should be investigated.

> I'm thinking that universities should be able to have sharing of information. So if one university subscribes, they should have an opportunity to make another university, a partner university have similar access to their resources, so we have like an open repository for universities where they actually subscribe together and then they can have access to similar resources. And like for each university or each research institute, some of which are very small, to actually subscribe to individual institutions, it becomes very expensive, I think. So if they can be able to share their networks, they can actually share access to the resources and then that can make it much better because in terms of universities, they are different in terms of sizes, in terms of budget and other things. So there's some smaller universities which cannot probably afford to subscribe to all the journals or other big publishers, but if a big university can subscribe, they can then provide access to a smaller university or smaller research institute so that the quality of access to library information can actually become better. I don't know if it works. (35-year-old male from Zimbabwe)

Summary and conclusions

The results presented in this chapter about research funding confirm the general picture that emerged in Section One of the book. There is a clear correlation between age and success in securing funding for research. The older (and more established) one is as a scientist, the more likely one is to secure funding. Although our respondents indicated that they are able to secure funding from both international and national sources, it is also clear that those who are more successful in doing so (those who indicated that they receive more than 50% of their funding from either source), secure significantly higher amounts of funding from international sources. In general, scientists in the fields of agricultural and health sciences have more access to funding and also to higher amounts of funding. The interviews highlighted the many barriers faced – especially by young scientists – in accessing and securing funding. Not surprisingly, then, they suggest that more support be given to them both by national agencies as well as within their own institutions in sourcing research funding.

CHAPTER 6

Lack of mentoring and support

Catherine Beaudry, Johann Mouton and Heidi Prozesky

Introduction

Van Balen et al. (2012) classify the crucial factors influencing scientific careers as individual, ascriptive (e.g. social background or family situation of respondents), structural (e.g. support by mentors, inclusion into networks, the impact of human-resource instruments of universities), as well as labour-market related. Their findings highlight that research performance – measured narrowly in terms of numbers of publications and citations – have little impact on the probability of individuals remaining in scientific careers. In fact, support by mentors and the inclusion in social networks have a substantially greater impact on careers. The authors conclude that there is not one discriminating factor that identifies which talents remain in academia, but rather an accumulation of advantages and disadvantages that leads to different careers (Van Balen et al. 2012). Friesenhahn and Beaudry (2014: 57) highlighted a number of challenges for early career scientists, in particular in Africa. Notably, they found that the 'mentoring and support structure' is perceived as vital for young scholars around the globe, including those from the African nations that participated in their survey (mainly Nigeria and South Africa).

Figure 40: **Disaggregation of career challenges, by age category**

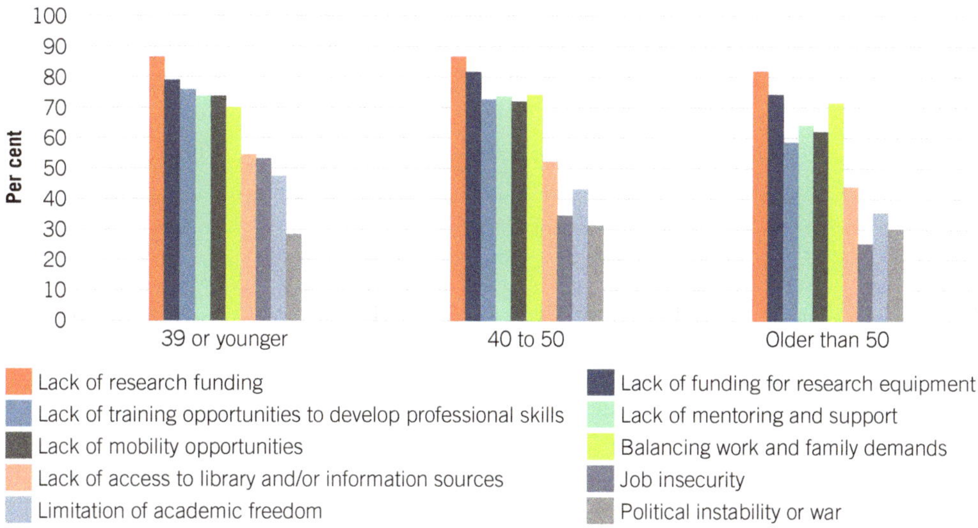

We have already discussed in the previous chapter the fact that (young) scientists consider lack of funding as the greatest challenge to their science careers. But it is interesting to note which challenges respondents rated as the third and fourth largest challenges.

In Figure 40 we compare the rankings of the three age cohorts on the list of stated challenges. Perhaps not surprisingly, the younger cohorts identified lack of training opportunities to develop their skills, as well as lack of mentoring and support, as their third and fourth largest challenges.

Mentoring received during career

Respondents were requested to report whether, during their career, they had received mentoring, support or training regarding seven aspects: career decisions, introduction to research networks, attaining a position/job, research methodology, fundraising, scientific writing, and presenting research results. In each case, respondents were provided with three response options, measuring two dimensions. First, whether mentoring, support or training had been received more than rarely; and secondly, the perceived value thereof.

Figure 41: **Percentages of young scientists indicating that they have never or rarely received mentoring on specific issue listed**

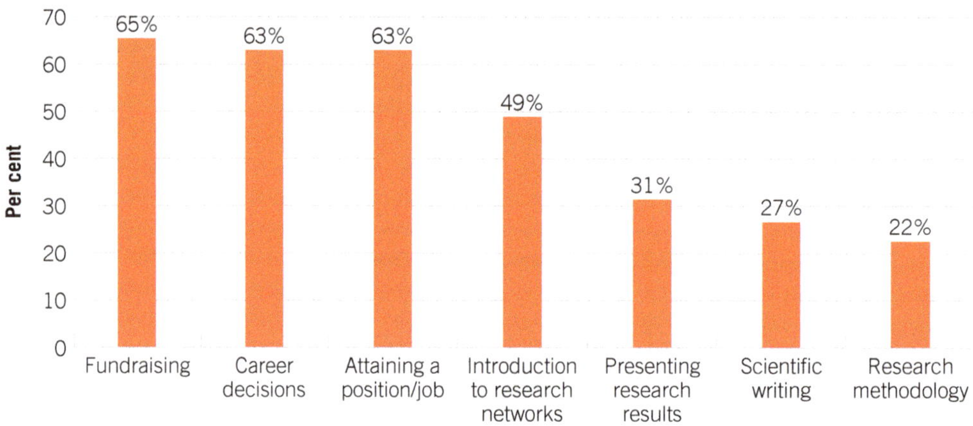

The results (Figure 41) are interesting, as the young scientists (39 years or younger) clearly indicated that advice on career-related decisions (including fundraising, attaining a job, and introduction to research networks) were the areas on which they did not receive, or very rarely received, advice. Advice that is more directly related to their research and doctoral studies (methodology, writing and presentation of research results) seems to have been more forthcoming. In the remainder of this section we present some of the ideas of our interviewees regarding these issues.

Fundraising

Young scientists need to know where, and how, to apply for research grants.

> I would love to be properly mentored on how to write research … To write for research grants and maybe do it practically with the mentors so that I'm finding out those things that are very pertinent for grant bodies that help them to transform … But, you know, there is no one to enlighten you on that. (40-year-old female from Nigeria)

> Well I'd say one area definitely is requesting funding, grants etc. because … it seems to be a whole science to actually be able to write these proposals and have them be successful. Also just maybe general project management. (32-year-old female from South Africa)

The absence of an adequate mentor to help identify possible sources for funding is particularly acute for young scientists entering into fields not yet established within their institutions.

> I was the first one to complete a master's in applied mathematics … And you find that those who are there, there is no mentoring initiative … Did they know something, did they know any funding opportunity, is there anything else you can do? (32-year-old male from Kenya)

Various interviewees indicated that they desire their institution to offer training and mentorship for fundraising.

> For the very first time I did apply for some funds … but I recently got a reply, it wasn't successful. I feel, yes it was a good project but I just feel maybe I didn't have enough training on how to approach it. So sometimes you have to go and learn yourself and do a research, you know, what to do and what not to do, but it is always nice if your institution could regularly organise programmes to train academic staff on how to go about such fundraising. (32-year-old male from Nigeria)

> The other areas I'm doing quite well, but this mentorship for fundraising, I think I have an issue there. I need very good mentorship on that area to be able to perform, yes. That is an area where I need good mentorship. (40-year-old male from Kenya)

Career guidance

Various interviewees underlined the lack of guidance during the early phase of their careers as young scientists. For example,

> The major challenges that I have encountered … as a young scientist who's just completed a PhD … about two years ago. And you are on your own. Especially the top scientists. Some scientists, you are not likely to get somebody who has maybe some personal love for you, just to help you, to mentor you. There's no mentorship plan. So, you try to struggle on your own throughout your career. That aspect of mentoring is lacking. (32-year-old male from Ghana)

Despite a clear career plan, another interviewee lamented wasted opportunities for making progress in his career, which he attributes to the lack of career guidance provided.

> You know, I had a pretty good idea of what the outcomes that I wanted to achieve were. I wanted to get my PhD and establish a research group and become a professor and I'm pursuing it in academia. It's kind of what I wanted to do. I just feel that, yes, it wasn't 100% clear to me that there were steps in the process that were not actually documented anywhere … I tend to focus on reading and the network and things myself, and so I will say that perhaps if there had been some kind of a mentorship situation it's likely that I wouldn't have understood what people have told me. So for instance at conferences, I've been to many conferences and I've only recently really understood that I should be introducing myself to people … I imagined that conferences were about going for the contents. So there is this kind of let's say tacit I think, and I think that's a lot more clear to many people than it was to me … Even just asking established researchers what advice would you give to young academics in the first year of their career and actually just track whether people are doing that. (39-year-old male from South Africa)

It is expected of mentors that they should act as intermediaries, introducing young scientists to senior researchers and helping them to establish their own research networks.

> See, that's what the absence of a mentor is, because if you've got a mentor in your field working with you, guiding you, you know, steering you in a direction of success, that, you know, that guidance helps you to tap into certain networks and helps you to establish networks. (33-year-old female from South Africa)

It is suggested that if mentoring is to be effective as a means to help persuade students to pursue a career in academia, it should start at the undergraduate level.

> There are very strong undergrads, when they write in the exams, when they contribute to classes, even when they speak, you know, you know this person will be good teacher in this field in the future. But some of them, without that mentoring, would just want to finish the programme and look for a job, you know. They don't want to further and it will take them another five, seven, eight years before they realise, wow, maybe this is what I should be doing,

> maybe I should go for my masters, you know. When you start mentoring them from that level they start thinking of it, they may then decide to go for a postgrad, immediately after school, decided, you know, because then they have that in the back of their mind that this is something [they] can do. So I think the earlier it starts the better. (32-year-old male from Nigeria)

The interviewees underlined a number of areas within which they desire input to help them develop their careers. A young scientist expressed the need for an explanation of what a career in research entails, and required assistance in deciphering what she perceived as cryptic academic and field-related jargon.

> I think for me it's just uncertainty of really what does a career in research entail? And I don't think that there's really time that's taken to actually sit and talk with you, about what are the options, what are the drawbacks, what are the pitfalls? And also not really spoken to me about, you know, where can you build from? Or, how can I put this, so it's almost like my training, my identity, was really ingrained in a professional identity. And then having to step out into being a researcher, what I didn't really feel like I knew what was going on. Even just broadly, like processes, what are the different processes that's involved in research? Like there's so many committees involved, and people use acronyms for things … I still don't understand what that means and what they do. And then just when you think you have a handle on it, you learn about a new committee and a new something that you need to know about, and it's just things to me about the whole time, like, what do they call it? Core? That they open a core to a grant or something like that. Just even the terminology, like opening a core to a grant, I don't know what that means. I have no idea what that means. I was like, what is that, does it … I kind of assumed, okay it probably means something like they're saying you can apply for it now. It's opening so you can apply for it now if you want to, or something like that. But some people that are ingrained in academia and people that are ingrained in research take those things for granted. Even just the language of it. (34-year-old female from South Africa)

Young scientists want a mentor to help them understand all the procedures and requirements related to their academic work.

> So, there is so much procedures that you need to follow in an academic career, and nobody actually tells you. So once you get appointed, nobody tells you, you know, this is how you find a journal, or this is how you go for a conference … Even something as simple as salaries, I didn't know that there were actual scales for salaries and so I actually found out now, that I am getting underpaid, and I think if you have a mentor, the mentor tells you, look at the salary scales for university, if you are appointed on this level, this is what you should be getting. So for me it's just, what I would expect from a mentor … (30-year-old female from South Africa)

There was also an expression of a need for continuous professional development.

> And also that person should be in a position to share with you, more or less, yes, that there is a recipe for success, the obstacles that one can come up with and also, I mean, all of us, we have ambitions, you are working towards promotion. How one can work yourself up towards promotion and increasing your research cohort plus output also. (44-year-old female from South Africa)

> Professional development is, like, okay, you know what, you need to finish ... You need to start writing applications. I'll look at your cover letter. What does your CV look like? Don't write your cover letter like this, write it like that. Apply here. Go to this conference here because, you know, you probably need a future employer or something like that. (35-year-old male from South Africa)

Mentors should also be available to give advice when a crossroad in a career is reached.

> And, you know, oftentimes at crossroads in my career I've been able to reach out to them, explain the difficulties or the challenges that I was having and then they were able to advise, based on their experiences, on what was the best path to follow. (40-year-old female from Nigeria)

An interviewee expressed the desire to receive training on career decisions and job acquisition.

> I think that it would be useful to have such training, career decisions and fundraising or how to get a job or practical things? Yes that would be the best. You know in the developed countries that is what they are doing so that is the reason why they are progressing more. (32-year-old male from Nigeria)

Another interviewee describes his inability to develop a career path in academia and attributes this to the lack of clear career paths available for young, aspiring academics.

> Failure to secure a permanent job in ... was one of the impediments I was talking about. Because ... usually now to be a lecturer they require that you must have a PhD. But then their methods of recruitment are a bit, you know, not very clear. ... They take long to make advertisements and getting a job there is actually a process, if I'm to call it. So, even up to now I'm still pushing that I get a permanent job there and get a salary. Because right now I'm still a part-time lecturer and time has gone. Because I graduated in 2014; this is 2018. ... You just keep applying and it's, you know, they say, you know, we are going to advertise; they don't advertise. Then somebody says there's no money in the university. So you find yourself having to run around now maybe in other universities, try to get a teaching job here and there, try to get consultancies, you know? So it is not easy in that sense. (38-year-old male from East Africa)

Apart from the research-related and fundraising skills discussed above, a number of additional skills and knowledge are sought from mentors to help young scientists fulfil a range of academic-related functions.

University processes

Young scientists want to be inducted into the research culture and understand the relevant university processes.

> Mentors, they play a pivotal role, right, from the university culture, research, teaching, and you look at what is expected. Of course, everyone knows what is expected, but the mentor is the person who actually takes you through the whole process until you are fully accustomised into the university's processes. (46-year-old male from South Africa)

This includes department-specific guidance.

> It would be good if there more general and practical issues in my department. So, I mentioned to my head of department and to the dean as to what should be done for a kind of mentoring system, where there is a new lecturer, there should be a person that is a mentor to the new lecturer just to be able to find their way in the department. There were sometimes where I would get to a lecture hall and I don't even have the password to start the computer in there. It was obviously the information was not given to me. What happened, every time information was given to me was only when I was facing the problem. Nothing was given to me in advance towards it. I was facing a problem and I would ask for help and I would be very busy. I should have known about this and I should have known about this. My first semester I can say was quite lonely because of all of these issues. There was not any mentoring system. (42-year-old male from South Africa)

Teaching-related activities

Young scientists seek preparation for undertaking lecturing duties.

> When you get appointed at an academic position and you have to lecture there's no preparation for that. It's like this is what you're teaching, just go for it kind. I find that whole thing a little funny. (36-year-old male from South Africa)

In this regard it may be useful acquire input and advice from various members of staff.

> Sometimes you have got, you have retired people in the department … There should be a team teaching with them. We need to know how do they teach, what are the best strategies? Tacit knowledge that they have about clicking and connecting with students, and the delivering of content in certain ways, it has to be evident. (48-year-old female from South Africa)

Interviewees expressed a desire to improve know-how on how to engage with industry.

> Well, yes I think, all through my research I think it would have been good if I had a better understanding of how to work with industry, maybe how to take ideas to research to actual development to things that people actually use, as opposed to it just being an academic exercise. I think being tenured that is now more important now, because I don't think we always have the luxury of just doing research for its own sake. (33-year-old male from Kenya)

Scientific writing

Apart from identifying where and how to publish, there is also a need for continuous feedback to help improve scientific writing abilities.

> You know, things I've been mentioning earlier. Writing, identifying a journal, submitting your paper to a journal. As I'm sitting in front of my computer now, if I had to have a first draft of a paper, I don't even know where to go to, to submit it. You know like basic things like that, yes. Reading, reading, even if it means reading two paragraphs. Giving me feedback – okay, this is a problem with your writing, you're writing long sentences. Or … you know? (38-year-old female from South Africa)

Research methodology (and publishing)

The need for mentoring to help develop research skills is emphasised.

> Coaching is about teaching young academics the skills and the competencies that go with conducting research ethically and professionally and in keeping with your niche areas. The vigour of research is based I think on one's understanding of research designs, research paradigms, the instrument design itself or the instrument selection. The procedure that goes with it in terms of differing from how other researchers have done their procedures, its sampling strategies, even choosing of your actual research population or unit of analysis. And the of course if we had to get into the actual analysis itself and then the ethical considerations, limitations and so on. I just think coaching where an experienced researcher and academic and the professor then mentors but at the same time direct skills and competencies (54-year-old female from South Africa)

Young scientists want to know where (and what) they can (and should) publish, as well as where to find required resources.

> Having a mentor that can advise on procedures and on things like where to publish or where to find resources. … Sit with me and discuss where a specific topic that I'm working on would, you know, which journal would be

> appropriate to publish about that topic or things like that, or someone who has experience at this university with how to access funding or who to talk to about procedures, so that has been really a significant help in knowing someone who knows the resources and the system and who are willing to give you advice and guide you. (29-year-old female from South Africa)

> There's a general lack of mentorship, that's the best way to describe it. I mean your more senior researchers, they assume that you know all about research and publishing and all of that. But you don't and then you often find that you identify journal, which you think would be suitable, only to find out later, no it's not really a credible journal or not a reputable journal or whatever the case. Yes, mentorship. (32-year-old male from South Africa)

Impact of lack of mentoring and support on career

Respondents were asked to indicate to what extent ('not at all'; 'to some extent'; or 'to a large extent') a lack of mentoring and support may have impacted negatively on their careers as academics or scientists. In order to facilitate comparisons between various subgroups of respondents, response categories were collapsed to create a binary variable consisting of the two categories (1) 'not at all'; and (2) 'at least to some extent' (a combination of 'to a large extent' and 'to some extent').

Nearly three-quarters of young scientists indicated that a lack of mentoring had a negative impact on their careers (Figure 40). Does this proportion vary across scientific fields? A disaggregation by scientific field (Figure 42) shows that slightly higher proportions of young scientists in the health and engineering sciences (two professional fields) reported that a lack of mentoring has had some negative impact on their careers.

Figure 42: **Lack of mentoring and support to at least some extent, by age, field and gender**

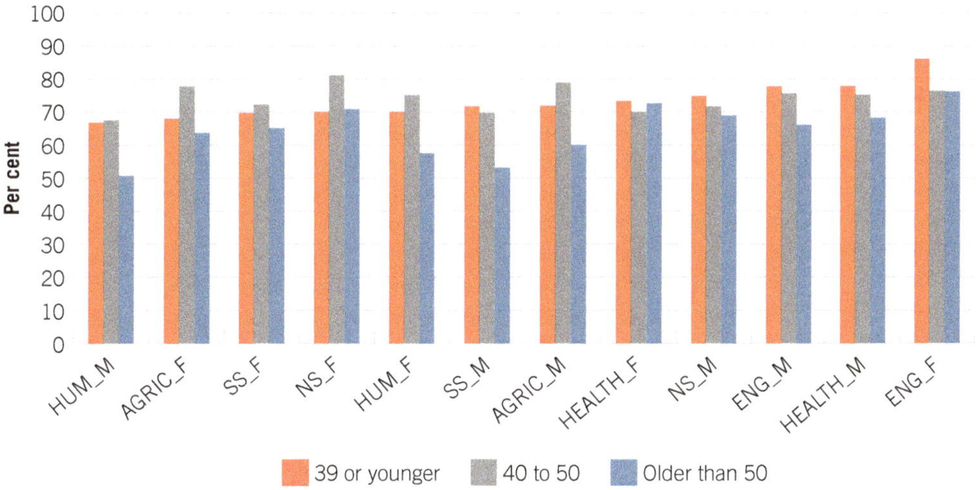

The qualitative interviews also showed that young scientists often struggle to find suitable mentors and role-models.

> So you can hardly find anybody that will be around, you have to push. Sometimes you push and you don't get anything and you tell yourself, if I continue to push I might not do the work I'm supposed to do ... That's what I feel. So there is a complete lack of any formal mentorship programme that can assist you in developing as an academic. (40-year-old female from Nigeria)

> In my eleven years here I've had no role-models ... Well, role-models are people that you look up to, I've looked up to a lot of people but not necessarily at this university. (38-year-old female from South Africa)

Despite the adoption of a 'publish or perish' culture at universities, support is not always forthcoming.

> The pressure, yes, there is pressure, but it is not very direct. Can I elaborate? What I mean, for example, to get a promotion, okay? You need to demonstrate that you have done research and you have published it, okay? But there is no special time or resource allocated to that. So it is up to you to decide to do research and get to the next level. To conform to the popular saying that you publish or perish. So it is actually some kind of paradox, you know? ... The universities do not facilitate their members to do research, but still they need you to do research. (37-year-old male from Uganda)

And where mentors were available, some interviewees pointed to the lack of experience of more senior staff as negatively impacting on mentorship initiatives. Insufficient numbers of established researchers at some institutions mitigate against effective mentoring of new academic staff.

> I needed to be mentored because I had just gotten my doctorate, but here am I now where there's nobody to mentor me. I was [the second-most-senior] scholar in that department because we are beginning that department. (40-year-old male from Kenya)

It is also difficult for young scientists in interdisciplinary fields to find mentors.

> So, the project and my research area is interdisciplinary and so while I have a supervisor in the one field and a supervisor in the other field, there's no one at UCT and also really in South Africa who I've met who is in my field and could be a mentor to me in my field ... There's no one directly in my field who knows more about the field than me, really. Because it's interdisciplinary and it's an emerging field it's very difficult to, you know, not have someone to learn from in your field. (28-year-old female from South Africa)

There is a perception among some interviewees that the more senior and experienced staff are disinclined to actively mentor young scientists.

> So a junior researcher requires mentorship or training in more locally focused research work that would be a bit of a challenge for them. But the exact opposite also happens sometimes that you find a lot of very senior academics that are not at all involved in research work, but are then heads of departments for institutions. And then they look at young and upcoming researchers as rogue scientists, people that are probably looking to do things the unconventional way. (35-year-old from Tanzania)

> These people who do these kinds of mentoring, they also do it in an informal way just when you're working with them. And they never seem to be the chairs of departments, the heads of schools, the dean or so on. They always seem to be just more colleagues. (30-year-old male from South Africa)

Local mentors purportedly lack experience, while international mentors lack incentive to provide support.

> Lack of mentoring locally, one, because most people were mostly inexperienced. We didn't have more qualified experts, senior experts. So, the result is mentoring problems ... And the second mentoring problem is, locally, for me and for other colleagues as an institute, as a whole understand, is that the mentors, the internationally recognised experts need some type of income in order to collaborate. (35-year-old male from Ethiopia)

The erosion of required experience and expertise at various institutions is attributed to 'brain drain'.

> So, if you look at the last 15 years at this university, we've lost 150 senior academic staff, so that's professor to associate professor level. I came in as a brand-new person to this university and I came in at the highest point in my department. That should never have happened, there should've been a professor or an associate professor above me. Now, where am I supposed to get academic mentoring from in that context and it's still happening ... We've got this knowledge gap that's starting to exist in the university where you've got newer scientists coming through and they've got no institutional knowledge, they've got no support structures from up above, it's really scary. (41-year-old female from South Africa)

Those mentors that do have experience are often too busy to mentor.

> Somebody is well-skilled but she doesn't have time to mentor the young scientist. This is a big challenge, not only in Kenya, I've experienced it in another East African country, I also experienced it in another South[ern] African

> country, that you do not have mentors, people who can show you something from the beginning to the very end … (35-year-old male from Kenya)

> Mentorship is also not adequate in our places because people, you find that they are busy. Maybe you have some senior staff, the university staff, but they are quite busy with other issues or they are quite busy with other students. So mentorship to young academics, stuff like that, maybe is not adequate … [There are] also quite a number of students, and therefore, he doesn't have much time for junior academics and stuff. (39-year-old male from Tanzania)

In other instances, mentoring is provided but is not effective. This may be attributed to a lack of commitment from mentors.

> Because some being other people who are senior, and they don't believe this person can contribute, or they know what they are doing, because they think they should bring someone onto the team who is not going to be able to contribute. They don't feel obliged to mentor people. So yes, it's bitter sweet. (34-year-old male from Uganda)

Some aspiring academics have become discouraged. A need for encouragement as opposed to negative feedback is sought from their mentors.

> I wouldn't say my supervisor is my mentor because he is so negative. I don't know, like I've submitted three or four times and there is no hope in his reports. There's no hope, you don't have that space of voicing and telling yourself that at least he liked something about my stuff. Every time you submit, it's just wait, wait, wait … I understand if you are a student, you have to get feedback, of course, but if you always get the negative feedback, what do you think? Are you going to be motivated? You would be demotivated, yes. (27-year-old male from South Africa)

> There's something on the term of academic bully. You know where almost you are, for lack of better word, discouraged from even starting, considering publishing because everyone will just tell you how difficult it is and not how everyone can crack and so on. You know and you usually hear this from people who have been in the game for a while and yet they themselves are doing it and you wonder but if you can do it, why can't I. You know it's those kinds of people that actually you are hoping would be the ones to encourage you to actually publish, you know like if I can do it, so can you … So, perhaps that's also a point to put in there, just academic bully you know. (24-year-old female from South Africa)

Impact of lack of training opportunities to develop professional skills

Respondents were also asked to indicate to what extent ('not at all'; 'to some extent'; or 'to a large extent') a lack of training opportunities to develop professional skills may have impacted negatively on their careers as academics or scientists. Again, in order to facilitate comparisons between various subgroups of respondents, response categories were collapsed to create a binary variable consisting of the two categories (1) 'not at all'; and (2) 'at least to some extent' (a combination of 'to a large extent' and 'to some extent'). For the entire sample, slightly more than two-thirds (69%) indicated that a lack of training opportunities has, to some extent, impacted negatively on their careers. Further disaggregation by field revealed huge differences (Table 20). Respondents in the humanities and social sciences seem to have access to more training opportunities than their counterparts in the STEM fields, as much smaller proportions of the latter identified a lack of training opportunities as a problem.

Table 20: **Lack of training opportunities, to at least some extent, to develop professional skills, by field**

	Humanities	Social sciences	Agricultural sciences	Engineering and applied technologies	Health sciences	Natural sciences	Total
N	108	543	388	375	772	974	**3 160**
%	43.9%	60.6%	70.5%	72.3%	73.4%	73.7%	**68.9%**

As Figure 43 shows, higher proportions of younger respondents (between 40 and 50, and 39 years or younger) were likely, across all scientific fields, to identify this as a challenge.

Figure 43: **Lack of training opportunities to develop professional skills, to at least to some extent, by age, field and gender**

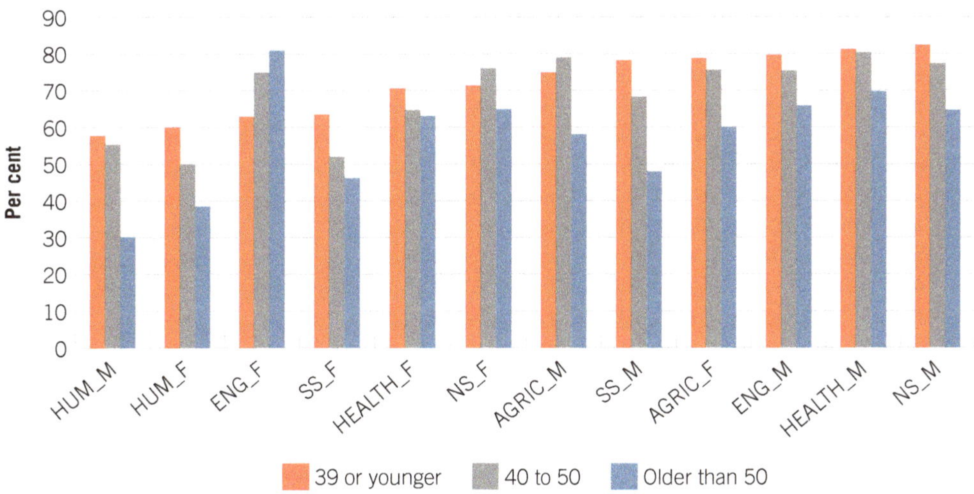

In conclusion, interviewees emphasise that their institutions should implement a clear and unambiguous mentoring system.

> There isn't a clear system on how to mentor, especially the young and when you come to research. There isn't a very clear-cut system you can say that this is how ... If you're interested in research, these are supposed to be your mentors or these are the people we suggest you be working with and they guide you ... There isn't a clear-cut mechanism on how to mentor people in research. I won't really say there is. (33-year-old male from Kenya)

This then necessitates the implementation of formal mentoring programmes targeting prospective, as well as new, academics.

> I think that if we want in the long term to have a growing research community, then we need to create a long-term programme that, you know, that looks at the people who complete PhDs, who have the potential to go on to become, you know, serious researchers, create opportunities for them to do post-docs, so allow them to go overseas. Create, at every stage, some form of programme, it doesn't have to necessarily be financial, but some form of either mentorship programme or encouragement programme, skills development programme that basically mentors and shepherds youngsters to this network so that they can become senior faculty at some stage. (35-year-old male from South Africa)

Summary and conclusions

Our findings correspond with the literature in terms of the important role mentoring and support, or rather the lack thereof, plays in the majority of young scientists' careers, especially in professional and interdisciplinary fields. For these young scientists, advice on career-related decisions were harder to come by, and the results of the lack of such advice more keenly felt, than training related to research. Where a lack of training opportunities to develop professional skills was highlighted, respondents in the STEM fields were more likely than their counterparts in the humanities and social sciences to experience this as a problem.

At the earlier stages of their research careers, young scientists voiced a need for more clarity on the requirements involved in pursuing a successful research career. These include an understanding of the relevant university cultures, preparation for undertaking lecturing duties, and navigating the intricacies involved in publication of research results in academic journals. At the later career stages, introduction to research networks and mentoring on fundraising are contributions more established scientists could offer. However, there seems to be an absence of experienced staff who are able to act as mentors and role-models, or a disinclination of available (often overburdened) senior staff to do so in an encouraging manner. It is therefore important that, aside from such individual mentors, the potential supportive function institutions may fulfil in many young scientists' careers was also highlighted in this chapter.

CHAPTER 7

Mobility and the careers of young scientists

Catherine Beaudry, Johann Mouton and Heidi Prozesky

Introduction

The (international) mobility of scholars and scientists is a key feature of the global science system (Knight 2008; Huang 2013; Rostan & Höhle 2014). Mobility is generally associated with positive effects for an institution (Welch 1997) and for the mobile individuals. Mobile researchers generally have a larger international network and perform better than their non-mobile peers (Cruz & Sanz 2010; Franzoni et al. 2012), they publish and are cited more often (Baruffaldi & Landoni 2012; Aksnes et al. 2013) and have better access to funding (Cañibano et al. 2008).

Given the specific nature of the mobility of many African scholars over the past four decades or so, and specifically the impact that the brain drain has had on knowledge production on the continent in the 1970s and 1980s, we decided to ask the survey respondents a number of questions on mobility. We first wanted to establish the extent of their mobility. We then asked them how important they regard mobility – i.e. being able to either work or study abroad – for their own career development. They also rated the working conditions abroad against their local working conditions. The latter theme was further explored during the interviews, during which respondents elaborated on the benefits of international visits and exchanges. The chapter concludes with a discussion on the relationship between mobility and the likelihood of obtaining research funding.

Recent mobility

Nearly 30% of the survey respondents indicated that they have studied or worked in a country other than their home country (i.e. abroad) over the preceding three years. The disaggregation by age shows that a higher proportion of respondents between 40 and 50 (40%) indicated that they had recently studied or worked abroad. The second-highest affirmative responses were recorded by the group 39 years or younger (37%), followed by 23% of the group older than 50.

The results of a disaggregation by scientific field are presented in Figure 44. The differences between the fields are statistically significant, and show that scholars in the humanities and agricultural sciences are most likely to have studied or worked abroad in recent years.

Figure 44: **Proportions of those who studied or worked abroad, by field**

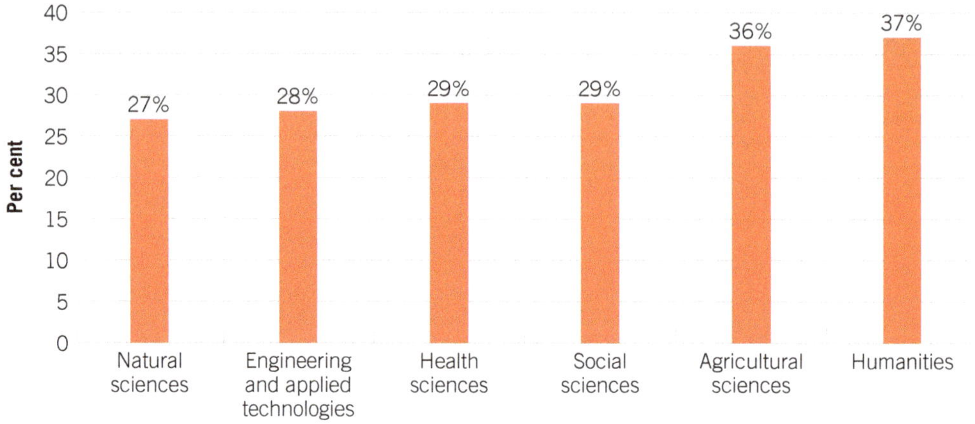

Because the dependent variable in these analyses is a dichotomous variable, we ran a cross-tabulation procedure with age, scientific field and gender as possible predictor variables to shed more light on patterns of mobility. The results are interesting (Figure 45). In almost all cases, age emerged to be a good predictor of whether a respondent had been abroad more recently. There is also a clear gender effect, with male respondents more likely to have travelled abroad.

Figure 45: **Age, gender and scientific field of internationally mobile respondents**

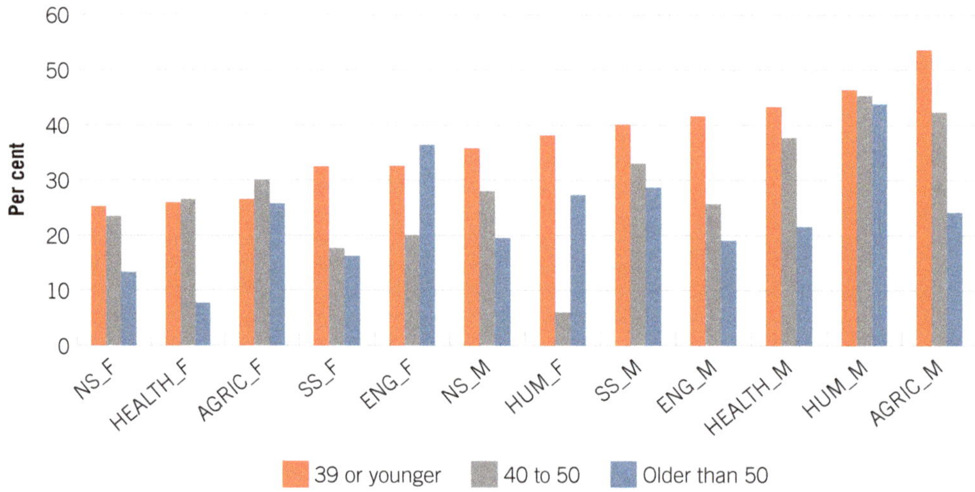

Of those who had worked or studied abroad, the majority indicated that this experience has been 'essential' (39%) or 'very important' (43%) for their career development (Figure 46).

Figure 46: **Rating of the importance of having studied/worked abroad for career development**

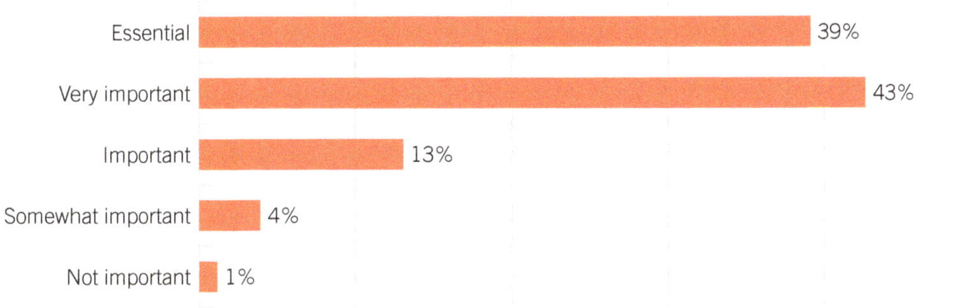

International-mobility characteristics differ according to sector of employment (Figure 47). The percentage of respondents who had been internationally mobile during the three years prior to the survey is proportionately the highest among those in international (81%) and non-governmental organisations (43%), followed by those in private research institutions (34%). Interestingly, only 33% and 29% of respondents in public research institutions and higher education institutions, respectively, indicated that they have travelled abroad in the preceding three years.

Figure 47: **International mobility according to sector of employment**

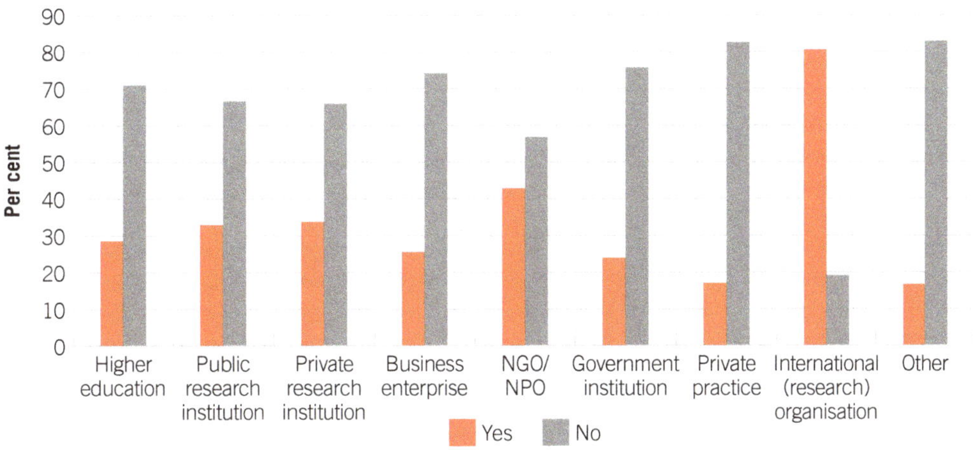

As far as region of residence is concerned (Figure 48), the percentage of respondents who had recently been abroad is the highest for respondents in Central Africa (47%) followed by near equal proportions of respondents from East Africa (38%) and West Africa (36%). Respondents from Southern Africa and North Africa are less likely to have travelled abroad in the preceding three years.

Figure 48: **International mobility according to region of residence**

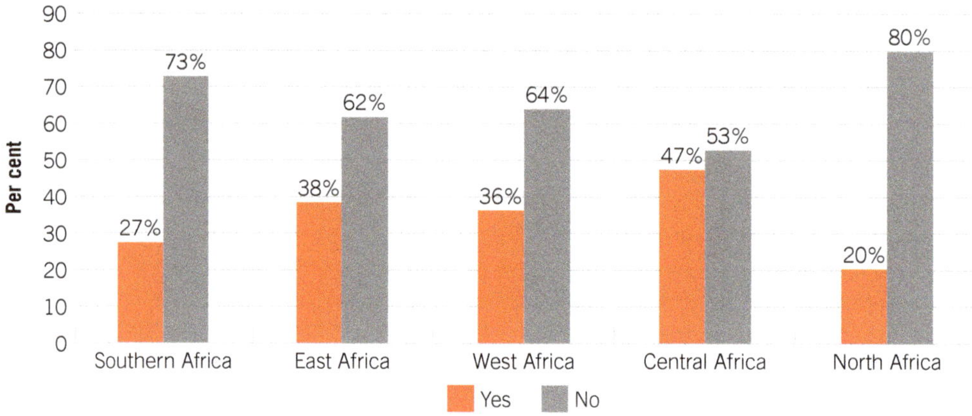

Benefits of international mobility

It is essentially the need for professional development that drives international mobility. This fact manifested in all of our interviews with young scientists. The quest for the acquisition of skills and knowledge unavailable in local institutions leads to the pursuit of studies abroad.

> Ah yes. At the time, there was no opportunity to do a doctorate in Benin. Years '98, '99, 2000, we did not have a doctoral school in Benin. The first doctoral schools date back to 2002/2003. To do a doctorate, you had to leave Benin. That was it. (Female respondent from West Africa)

> When it was time for me to do a doctorate for me in Belgium, there was not yet in Burkina. I did a doctorate in public health sciences. In Burkina Faso there were none, neither in sociology nor in public health sciences. (Female respondent from West Africa)

A number of additional benefits of studying abroad are described. These include the quality of the education and the prestige of the degree.

> I do not know if it's in the development of the career but often here people think that the one who studied in the North has better mastered this thing than the one who studied in the South. In recruitment, when there is a competition for recruitment, there are some who have a tendency towards the diploma of the North than towards that of the South. When we look at someone's resume, we say he did Canada, he did the United States, he did France. It sounds better than someone who did Dakar, Abidjan or Ouagadougou. So in the job market, those who have been to the North are selling relatively better than those who have made 100% South. (Male respondent from West Africa)

> The quality, of course, is better the other side. And I think my PhD is more recognised than the Ugandan one. I think it is taken to be a European PhD. (38-year-old male from Uganda)

> I can explain a little bit about that. I was a bit fortunate that when I did my masters it was in Europe, so in terms of the quality of the education, it was much higher in terms of publication and research, we did a lot of skills. (35-year-old male from Zimbabwe)

Individuals who have studied abroad are highly regarded, as it is assumed that they have acquired superior experience and expertise.

> We who have studied abroad, we are considered as people who have a great intellectual background. I play the role of leader because of that. I am a leader because I travelled the whole world we will say it. I went to Canada, I know the United States, France, I know Germany, so I know how it works in those countries. (Female from West Africa)

There are also abundant training opportunities provided at international institutions.

> I benefitted from [the] training when I was in Germany … You were offered one week, two weeks, three weeks' training on proposal writing, funding applications. How we can interact with funding organisations. Yes, how to prepare all those things, on all that I benefitted while I was in Germany … Yes, there are special models and actually it was always there I found there was a time table and different schedules for that. (39-year-old male from Tanzania)

Perhaps the most substantial benefit derived from having studied or worked abroad is the acquisition of research networks that may, in turn, lead to participation in research projects and funding opportunities.

> And, of course, you know, the facilities there were fantastic; the opportunities that I got, the connections that I got, that I made there with my supervisor and other people in Ireland enabled me to get now onto another project which is going on, which I'm doing as part of my kind of post-doc. With the same scientists that actually were in charge of that project where we did our PhD. (38-year-old male from Uganda)

> The advantage of studying in the North, it allows you to make you a network of friends in the North. When you study in the North, you can make a network of friends even if you go back to the South, you can stick to this network to have North–South funding. But if you stay studying exclusively in the South, apart from the fact that with publications, you can contact some authors who are in the same theme as you and that relationships can arise, it is difficult for

> people who study exclusively in the South to become partners of the North. (Male from West Africa)

> I studied in Germany and yes I was at one institute … And through that centre actually we were three; we were interacting with different working groups. Through those, the leaders of those working groups, we were able to interact with other institutes within Germany and outside Germany. And they were involving us in their research projects. Even they were involving us in preparation of their research proposals and we were getting exposure. (39-year-old male from Tanzania)

> The opportunities are a lot better than in Nigeria, but if you look at research funding, of course, before that time, I didn't get any research funding in Nigeria to carry out any of my research. But of course, I now got funding to carry out my research in Germany. And of course, I could see that while I was there, I could see that there [are] several other opportunities I could even apply for … So yes, you have more research funding opportunities. (35-year-old female from Nigeria)

International work and study allows scientists in Africa to access to expertise in fields that are not available locally.

> Some of the things which you would like to pursue, you may not find somebody who has those skills and experience. That is one. So, when you are looking for an opportunity, for example, who will train me on this, sometimes within the country you may not find one and therefore you are supposed to look for such opportunities out of your country. So, that is where it is a big challenge, especially lack of training opportunities. For you to apply outside the country, you must wait until there is a call for a given kind of skill or some people looking for something for you to go and train. So, sometimes it may not come along the area of your interest and therefore that is where it is lacking. (35-year-old male from Kenya)

Further benefits of international work and study include exposure to individuals that are considered to be the top experts in their respective fields.

> Also the opportunity, you know, you have to meet people who are at the top of your speciality. You see some of the books we read in Africa here, when you go to the UK you'll be interacting with the very people who are writing those books. So, it's a great opportunity going to the UK to do a PhD. (38-year-old male from Nigeria)

An interviewee from West Africa feels that there are too many distractions at home, which generate the need to work and study outside their home country.

> Well, you know there are the next-of-kin, the friends, so sometimes you get distracted. Sometimes you move away from your career, from research. You really have to get out of the country or the city to concentrate so you can really get the results. That's what I've always done. Since my return to the country in 2004, I have always planned, not even a month or two months of post-doc in Germany with my teacher who supervised me. This allowed me to move forward and this often allowed me to effectively achieve results. Otherwise in Benin it's really difficult. (Male from West Africa)

Access to books and office resources are emphasised.

> You have more research resources, especially in terms of scientific literature. In Germany, I had access to so many books. In Nigeria, the highest number of books I could collect from the library at the time is four, I could only collect four books … but in Germany, I had access to so many books. So many. In fact, I could have as many as 20, 30 books at a time, and at the time, you have a lot of scientific literature, current scientific literature. And literature that is specific to the field. Most of the books we had here are general books to a field. For example, let's say pragmatics or linguistics, things that are general. But then when you're looking for specific areas, you'd still find that you don't have access to them. So I had a lot of research resources in Germany. (35-year-old female from Nigeria)

> For example, I had an office space, we had a computer with the internet, and we had a telephone. We had this office material, we had paper we could print, we could scan, no questions. I don't have a scanner, if I want to scan you can ask can you scan that. [In Tanzania] I have to leave the campus here to go to town to scan. Yes, these are the things that really they restrict someone to be active. (39-year-old male from Tanzania)

Thus, many interviewees referred to the various benefits of being mobile, one of which is the access to more funding. In the questionnaire, the section on funding asked our respondents to indicate whether they had received any funding in the preceding three years, and if so, whether they were the primary recipients of such funding. In order to 'test' the hypothesis that greater mobility correlates with greater access to funding, we subsequently compared respondents who have travelled abroad and those who have not, in terms of their responses on the receipt-of-funding question. The results presented in Figure 49 clearly show that one is more likely to have received funding if one had previously worked or studied abroad (taking into account that the majority of the young scientists have not studied or worked abroad).

Figure 49: **International mobility according to receipt of funding (young scientists only)**

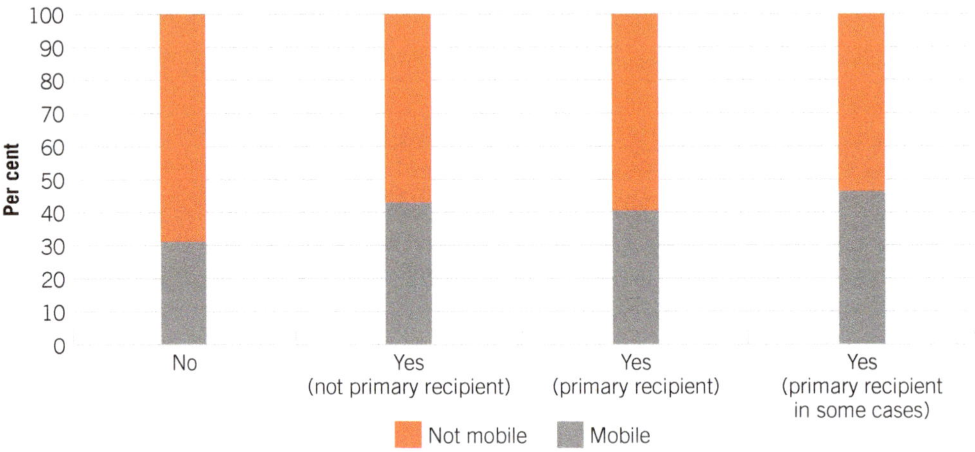

Perhaps even more telling is the relationship between having studied or worked abroad and the proportion of funding secured from international sources. Conducting an ANOVA (controlling for mobility, age, gender and field), clearly shows the benefits of mobility in terms of securing international funding. The results presented in Figure 50 show that age and field matter in securing international funding. In general, the older a respondent is, and if he/she works in the natural and health sciences, the more likely he/she is to have received funding in the previous three years. But when one controls for international mobility, the results show that the mobile scientists in all fields are more likely to have secured international funding. The graph also shows that young scientists who are mobile are equally (and in some case more) likely to secure international funding.

Figure 50: **Reported percentage of funding from international sources, by age, field and mobility**

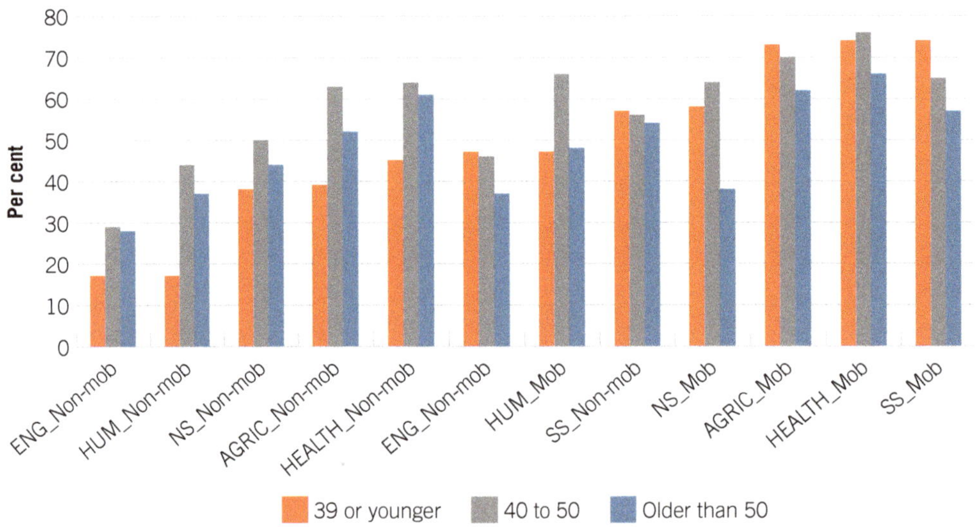

Comparison of study/working conditions abroad to those in home country

We also asked the internationally mobile respondents to compare the study or working conditions in their home country with the ones they had experienced abroad. We proposed six elements: (1) employment or job security, (2) work–family balance, (3) training opportunities, (4) opportunities for research collaboration, (5) research resources (personnel, scientific literature, material, etc.), and (6) research funding opportunities.

The results (Figure 51) are as one would expect. The country abroad is rated as better than the home country in terms of research resources, opportunities for collaboration, funding, and training opportunities.

Figure 51: **Rating of study/working conditions abroad compared to those in home country**

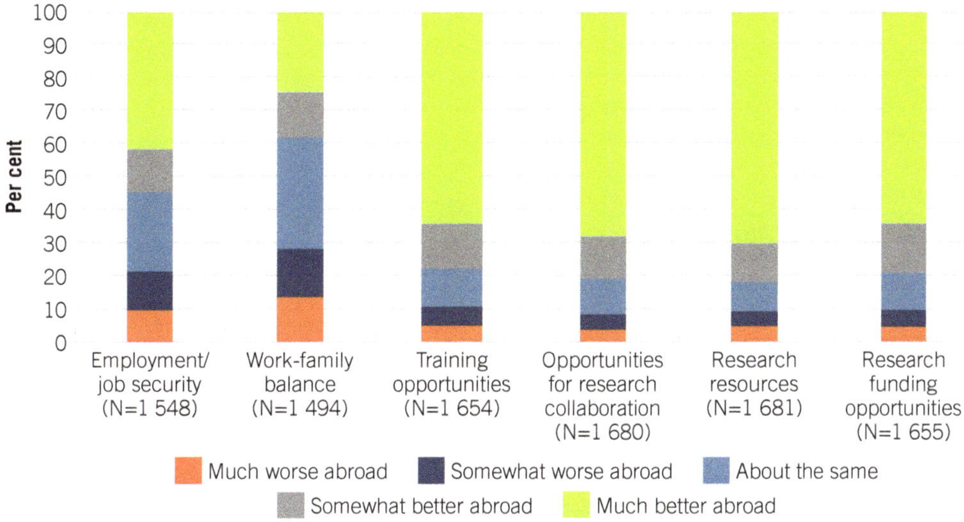

Given the very positive ratings of working and studying abroad, and specifically with regard to various research career and professional-development considerations, the results of the disaggregation (Table 21) of these elements by age are not surprising. In most cases, younger scientists are more inclined to report the advantages of studying and working abroad than their older counterparts are. The huge gap between the study and working conditions locally and internationally is perhaps best illustrated by the huge proportions of young scientists who consider working and studying abroad as better in terms of access to research resources (91%), as well as opportunities for research collaboration (88%), training (87%), and research funding (85%).

Table 21: **Rating of study/working conditions abroad, by age category and different factors**

	Worse abroad	About the same	Better abroad
Research resources (personnel, scientific literature, material, etc.)			
39 or younger	3%	6%	91%
40–50	6%	6%	89%
Older than 50	7%	10%	83%
Opportunities for research collaboration			
39 or younger	4%	8%	88%
40–50	6%	8%	86%
Older than 50	6%	13%	82%
Training opportunities			
39 or younger	5%	8%	87%
40–50	8%	8%	84%
Older than 50	8%	16%	76%
Research funding opportunities			
39 or younger	6%	8%	85%
40–50	6%	9%	85%
Older than 50	8%	15%	77%
Employment/job security			
39 or younger	17%	23%	60%
40–50	19%	25%	56%
Older than 50	23%	32%	45%
Work-family balance			
39 or younger	28%	33%	40%
40–50	27%	31%	42%
Older than 50	25%	43%	32%

During the interviews, the interviewees elaborated on the numerous benefits of international study and travel. Marked differences between local and international research environments are described. For example, international institutions are perceived as providing better funding and collaboration opportunities.

> I'm planning to go and join a research institute in Germany, because I believe that when I go there for the three years if offered, I'll be able to develop my research career better than actually when I'm here in Africa. So I want to go there and collaborate with international researchers as well as do it in an institution where there is funding for both research and for publications, where the environment is more conducive for research than it is here in Africa. (35-year-old from Zimbabwe)

Interviewees also consider education and training opportunities at home to be inferior when compared with those at overseas institutions. This is attributed to higher levels of expertise and a greater concentration of experience abroad.

> When I did my masters it was in Europe, so in terms of the quality of the education, it was much higher in terms of publication and research, we did a lot of skills. But when I come back home, my teamwork, my colleagues and other people I work with, they don't have sufficient skills and training, particularly in terms of scientific writing and publication. So what then happens is they kind of, within university they publish those journals and the quality of the research is not good enough to get into. So when you're surrounded by such kind of people you are pulled down in terms of quality, in terms of achieving your personal objectives. (35-year-old male from Zimbabwe)

> Because in general in Nigeria I learnt on my own, I never, nobody taught me ... In Germany on the physics course I was actually led through the process of publishing ... I also had contact with a lot of experts, a lot of experts in different areas that can help you ... support that helps you work better and do your work in time. Sometimes in Nigeria if you want to do something ... you really have to do that on your own. (35-year-old male from Nigeria)

Unfortunately, the expertise and experience are not transmitted through collaboration with African institutions.

> My personal feeling is that we have a lot of people in Europe and US who are doing research on Africa. But if you look at their publications, they don't include Africans ... there's not that mentorship. They should probably have taken one African to become the fourth author or some part of the team and then they will be learning through that process. So there's that divide in terms of the quality of research, so there's no proper mentoring and support for African researchers, because if you work with them, there's the experience, they have the resources. (35-year-old male from Zimbabwe)

This discrepancy between conditions overseas and at local institutions has resulted in significant erosion of local expertise.

> What I've discovered, like in my case ... we had this exodus of qualified lecturers ... of PhDs and professors. So when they went outside the country, mostly United States, South Africa, the UK, you tend to remain with people who are not quite qualified to do research ... So you find you remain with somebody who hasn't published a paper, does not even know how to supervise a master's degree or a PhD student, in terms of finding something that is new, at a master's level, basically, you don't need to just regurgitate what is there. You have to come out with new solutions, new ideas. So you find quite often ... you find a lot of guys had gone to other countries. (35-year-old male from Zimbabwe)

> Sometimes when people in the North are part of this mobility, it's a brain drain. Sometimes it is a loss of grey matter for Africa. Here you are in Canada.

> Maybe Senegal has lost you temporarily or permanently ... The Canadians are very strong in that. Sometimes you are allowed to go with your family and after, madam and the kids do not want to go back (laughs). It's on purpose. (Male from West Africa)

Brain gain

On a more positive note, a major advantage of mobility for home institutions, in the case of returning researchers, is gained through the resultant brain gain.

> Actually it's not leaving for good but leaving only for opportunity and then coming back because this is my homeland, it's where my family is ... [The reason I am going is] to expand the scope of my work and also to be able to practice what I have, if I had the opportunity ... It's not like I'm migrating, I'm not migrating. (39-year-old male from Tanzania)

> I always view my publications as being done growing crops in a desert, you know. They can grow if you can get something, but it's under very difficult conditions, so I've always wanted to go to a more conducive environment, and then I'm sure I'll be able to do even much more than I'm able to do when I'm here in Africa. So in short I'm saying I'll be going to Germany to join a research institute so that I can develop my research career, and then hopefully after the three-year contract I'll come and also build capacity of many African researchers so that we can strengthen research in Africa. (35-year-old male from Zimbabwe)

An interviewee from Nigeria expressed his loyalty to his local institution as a result of continued support to develop him professionally.

> When I finished my first degree, like I said, after my service I was retained by my university. My university allowed me to go and do a master's degree. Because, see, not all universities will do that. They were still paying my salary while I was outside ... When I came back, the rule is that when you come back you are supposed to serve the university. If you're still outside for one year you need to serve the university for two years. If you stay outside for two you'll have to serve the university for four years ... So I went straight one year, came back in 2007, went back in 2008. You can see the university didn't complain that I ... Why not stay until you serve your two years? The university said, no, no, no, go, go, go. I still went, spent three years. So I felt that the university contributed to who I am today. So I felt I definitely gave them value for their money ... So that was why I decided to stay back and then see how I can transfer my knowledge up to the society. (40-year-old male from Nigeria)

Considered leaving one's country

Viewed against the background of the results of the previous sections, it is not surprising that large proportions of our respondents indicated that they have considered leaving the African country where they were working or residing at the time of the survey: 20% said that they have 'often' considered doing so, whilst a further 51% indicated that they 'sometimes' think of leaving their home country. For the remaining minority (29%), this has never been a consideration.

The disaggregation by age category (Figure 52) shows small but interesting differences, with nearly 80% of all young scientists indicating that they either often or sometimes consider leaving the country where they work/reside.

Figure 52: **Considering leaving the country where you work/reside, by age category**

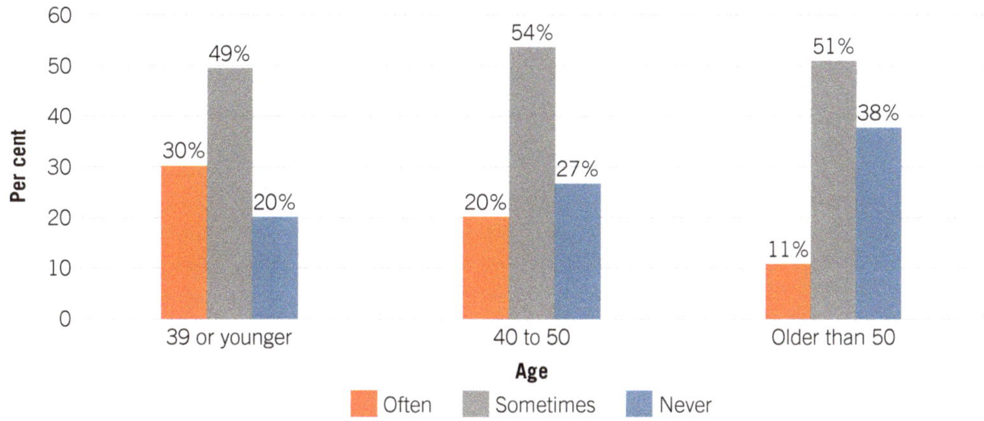

An open-ended question asked respondents to provide the three main reasons why they would leave the country where they work. We constructed various categories to regroup all answers, and to identify the 20 most-frequent first, second and third reasons mentioned (see Table 22).

We find both push factors (features of the country where they work) and pull factors (attractive features of a foreign country) driving respondents' willingness to leave the country where they work. Some respondents also stressed their homesickness. There is a variety of categories, referring to both academic and non-academic dimensions. We can see that the three most-frequently cited first, second and third reasons concern career prospects, academia, and general quality of life, respectively.

Table 22: **First 20 reasons to leave African country**

First reason	N	Second reason	N	Third reason	N
Career prospects/Advancement of career/Job opportunities/Job security	398	Academic reasons	418	Health care/Social amenities/quality of life/ Family/General Infrastructure	179
Politics/Economy/Social climate	375	Career prospects/Advancement of career/Job opportunities/Job security	292	Salary/Income/Revenue	145
Resources/Equipment/Facilities/Infrastructure	273	Salary/Income/Revenue	231	Resources/Equipment/Facilities/Infrastructure	131
Salary/Income/Revenue	270	Funding	205	Career prospects/Advancement of career/Job opportunities/Job security	124
Insecurity/Crime/War/Instability	221	Social welfare and state provision (Education/ healthcare/family security/quality of life/ infrastructure – country)	204	Insecurity/Crime/War/Instability	114
Funding	203	Insecurity/Crime/War/Instability	175	Administration/Bureaucracy/System/Corruption/ General governance/Research policy	99
Education/Training/Mentoring/Studies	193	Institutional reasons (Administration/ bureaucracy/efficiency/corruption/infra-structure – institution)	173	Funding	97
Working conditions/Environment	179	Politics/Economy/Social climate	101	Politics/Economy/Social climate	85
Health care/Social amenities/Quality of life/ Family/General Infrastructures	166	Acquire new skills/knowledge/expertise/ experience/development	83	Collaborations/Networking	75
Better opportunities/Greener pastures/Better prospects for the future	157	Political/social/economic climate of country and/or institution (Protests/weak currency/ failing economy/limited market/barriers to private sector entry)	79	Working conditions/Environment	61
Acquire new skills/knowledge/expertise/ experience	140	Resources/Equipment/Facilities/Infrastructure	79	Education/Training/Mentoring/Studies	54
Research	129	Personal security (crime/war/instability/fear for well-being of self and family)	73	Better opportunities/Greener pastures/Better prospects for the future	53
Racism/Affirmation action/Discrimination/ Xenophobia	74	Career opportunities (Employment prospects/ working conditions/mobility)	59	Acquire new skills/knowledge/expertise/ experience	51
Administration/Bureaucracy/System/Corruption/ General governance/Research policy	56	Working conditions/Environment	46	Racism/Affirmation action/Discrimination/ Xenophobia	45

First reason	N	Second reason	N	Third reason	N
Collaborations/Networking	49	Racism/Affirmation action/Discrimination/ Xenophobia	45	Research	43
Academic quality/Higher level of research	46	Adventure/Challenge/Experience with other cultures and society/Change	30	Adventure/Challenge/Experience with other cultures and society/Change	26
Post-doctorate position	42	Collaborations/Networking	21	Personal	25
Motivation/Job satisfaction	41	Better opportunities/Greener pastures/Better prospects for the future	18	Recognition	21
Recognition	39	Discrimination (racism/xenophobia/affirmative action)	16	Motivation/Job satisfaction	16
Exposure/Visibility/Impact	31	Conferences	14	Conferences	14

Table 23: **Reasons to leave African country where one works**

First reason	N	%	Second reason	N	%	Third reason	N	%
Career opportunities (Employment prospects/working conditions /mobility)	836	25	Academic reasons (Freedom/ Collaboration/mentoring and support/ recognition /conferences/visibility/ impact)	472	18	Institutional reasons (Administration/ bureaucracy/efficiency/corruption/ infrastructure – institution)	260	15
Political/social/economic climate of country and/or institution (Protests/ weak currency & economy/limited market/barrier	400	12	Career opportunities (Employment prospects/working conditions /mobility)	421	16	Career opportunities (Employment prospects/working conditions /mobility)	258	15
Further studies/training/acquire new skills/expertise/experience	366	11	Institutional reasons (Administration/ bureaucracy/efficiency/corruption/ infrastructure – institution)	255	10	Academic reasons (Freedom/ Collaboration/mentoring and support/ recognition /conferences/visibility/ impact)	252	15
Academic reasons (Freedom/ Collaboration/mentoring and support/ recognition /conferences/visibility/ impact)	365	11	Personal security (crime/war/instability/ fear for well-being of self and family)	250	10	Social welfare and state provision (Education/healthcare/family security/ quality of life/infrastructure – country)	179	11

First reason	N	%	Second reason	N	%	Third reason	N	%
Institutional reasons (Administration/ bureaucracy/efficiency/corruption/ infrastructure – institution)	335	10	Salary/remuneration	232	9	Salary/remuneration	149	9
Salary/remuneration	270	8	Social welfare and state provision (Education/healthcare/family security/ quality of life/infrastructure – country)	205	8	Personal security (crime/war/instability/ fear for well-being of self and family)	114	7
Funding	226	7	Funding	205	8	Political/social/economic climate of country and/or institution (Protests/ weak currency & economy/limited market/barrier	114	7
Personal security (crime/war/instability/ fear for well-being of self and family)	221	7	Political/social/economic climate of country and/or institution (Protests/ weak currency & economy/limited market/barrier	190	7	Further studies/training/acquire new skills/expertise/experience	108	6
Social welfare and state provision (Education/healthcare/family security/ quality of life/infrastructure – country)	166	5	Other	110	4	Funding	97	6
Discrimination (racisms/xenophobia/ affirmative action)	74	2	Further studies/training/acquire new skills/expertise/experience	101	4	Discrimination (racisms/xenophobia/ affirmative action)	52	3
Return to home country (Contract expire /homesick/limited opportunities)	34	1	Discrimination (racisms/xenophobia/ affirmative action)	76	3	Other	47	3
Adventure/Challenge/Experience with other cultures and society/Change	28	1	Adventure/Challenge/Experience with other cultures and society/Change	30	1	Adventure/Challenge/Experience with other cultures and society/Change	26	2
Other	23	1	Cultural issues/language barriers	11	0	Return to home country (Contract expire /homesick/limited opportunities)	14	1
Immigration issues (Visa/working permit/partner and family relocation)	9	0	Immigration issues (Visa/working permit/partner and family relocation)	7	0	Cultural issues/language barriers	8	0
Cultural issues/language barriers	4	0	Return to home country (Contract expire /homesick/limited opportunities)	7	0	Immigration issues (Visa/working permit/partner and family relocation)	1	0
Total	**3 357**	**100**	**Total**	**2 572**	**100**	**Total**	**1 679**	**100**

Lack of mobility opportunities

Our discussion thus far has focused on the relatively mobile group of young scientists in Africa and their perceptions and experiences of the benefits associated with having been able to study and or work abroad. But a large proportion of our respondents clearly have not had the benefit of either studying or working abroad. All respondents were thus asked to indicate to what extent ('not at all'; 'to some extent'; or 'to a large extent') a lack of mobility opportunities may have impacted negatively on their careers as academics or scientists.

In order to facilitate comparisons between various subgroups of respondents, response categories were collapsed to create a binary variable consisting of the two categories (1) 'not at all'; and (2) 'at least to some extent' (a combination of 'to a large extent' and 'to some extent'). Respondents of all ages indicated that a lack of mobility opportunities has impacted – at least to some extent – negatively on their careers. This result applies specifically to the younger age cohorts (Figure 53).

Figure 53: **Impact of lack of mobility opportunities, by age category**

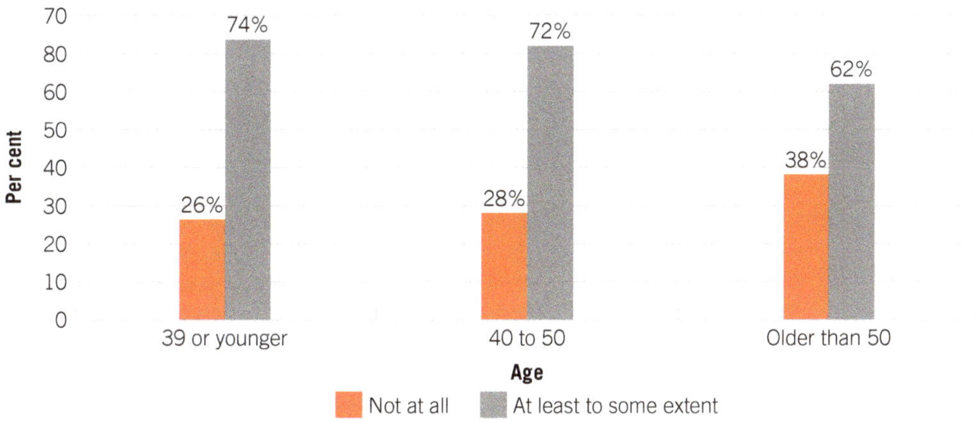

A comparison between respondents in various scientific fields in terms of their perceived negative impact of a lack of mobility opportunities (Table 24), shows some differences. The most notable result is the fact that a lack of mobility seems to have had the least negative impact on scholars in the humanities and social sciences. One would assume that the access to state-of-the-art equipment and laboratories is the main reason why higher proportions of respondents in the natural and engineering sciences indicated that not having had the opportunity to study or work abroad has had some negative impact on their careers.

Table 24: **Lack of mobility opportunities, by field**

Negative impact of lack of mobility opportunities		Field						Total
		Humanities	Social sciences	Health sciences	Agricultural sciences	Engineering and applied technologies	Natural sciences	
Not at all	N	102	347	346	146	143	320	**1 404**
	%	41.0%	39.7%	33.5%	27.4%	26.7%	24.5%	**31.0%**
At least to some extent	N	147	526	688	386	393	987	**3 127**
	%	59.0%	60.3%	66.5%	72.6%	73.3%	75.5%	**69.0%**

During the interviews we probed on the challenges young scientists face to gain the opportunity to go abroad. In one instance, an interviewee feels opportunities to travel may be trivially denied by their current employer.

> You know we are employed here and everything we do here we have to get permission from our superior. But if I say I want to travel to South Africa I have to write a letter to request it, for permission to travel. And you know it will [be] within the power of my superior to say yes you should go or you shouldn't. Even without scrutinising what am I going to do there? Is this work for my career or is it not? If they say no and then I travel then they will expel me from the job. (39-year-old male from Tanzania)

A lack of information regarding opportunities abroad is also cited as a factor inhibiting mobility.

> I feel like a typical academic should have a cross-breed of different orientations. I feel so strongly, yes ... the issue is information asymmetry. Like I think there are or the extent to which we're aware, I think, is really limited. I think so, yes. (32-year-old male from Nigeria)

A number of interviewees lament their inability to attend international conferences primarily because of funding restraints.

> I have been accepted for conferences in Barcelona, in the UK but due to the shortage of funds I couldn't make it ... These are very serious challenges. Like the one in Barcelona is a workshop on dry lands and it would have helped me greatly because there were scholars from Italy that have worked in the desert of Namibia and very similar conditions to where I work in the northern part of Nigeria, in the workshop. And it was a free workshop, the only thing I needed was to get the transport and the accommodation and that's all, so I couldn't make it because of a shortage of funds. (37-year-old male from Nigeria)

> I have been able, but most of them locally. For a while, I haven't gone for an international one for out of the country. But locally, yes ... Most of the time it's the funding because the university doesn't afford the international

> conference, so most of the time, it actually supports the local one. So for the international one, I have to look for the funds out there. So if I'm not lucky, I won't be able to attend. (33-year-old male from Kenya)

Summary and conclusions

Being able to travel abroad and engage with like-minded scholars is an integral part of being a scientist. Those scientists who are more mobile gain huge advantage in accessing international networks and the top researchers in a specific field, in finding out about new funding and collaborative opportunities, and in gaining access to work in the best laboratories in the world with state-of-the-art equipment. These are the building blocks of the academic capital that established scientists and scholars have built up over the span of their careers.

This chapter has shown how scientists of all ages and fields value being able to study and work abroad. Although a substantial proportion of the young scientists have had the opportunity to do so in the preceding three years (40%), it is still noteworthy that the majority has not had this opportunity. Those who have benefitted from international visits have listed the major benefits to them and their careers.

On the negative side, those who have not had these opportunities have voiced their concerns about the lack of being internationally mobile. They also lament the fact that they don't always know about such opportunities and even in cases where they wish to travel overseas some institutional barriers prevent them from doing so.

PART THREE

Research performance

CHAPTER 8

Research publications

Johann Mouton and Heidi Prozesky

Introduction

This chapter is organised under two main headings. First, a discussion of the survey results regarding research publications and specifically the relationship between age, scientific field and gender of the researcher; and second, a discussion of the factors that respondents listed as impacting either positively or negatively on their research productivity. The latter discussion mainly draws from the qualitative interviews that were conducted with respondents following the survey.

The relationship between the age of researchers and their scientific production has been an object of enquiry for some years now (for a review of the topic, see Feist [2006]). Lehman (1953) demonstrated that major contributions are likely to occur when scientists are in their late 30s or early 40s, and thereafter decline rapidly. Since this seminal paper, the research on this issue has diverged. Some studies have claimed that scientists conduct their best work while young (Einstein, Newton and Gauss are obvious examples), while others argue that knowledge matures with age (Plank, Braun and Cram were in their 40s when they formulated their theories).

The first group of studies advocates that younger researchers are more productive and more likely to be cited than their older colleagues (Gieryn 1981; Horner et al. 1986; Over 1988), and that extraordinary achievements tend to occur before the age of 40 (Adams 1946; Zuckerman & Merton 1973; Zuckerman 1977; Stern 1978; Simonton 1994; Dietrich & Srinivasan 2007). This tends to support Kuhn's (1962) argument that young researchers bring a 'fresher perspective' on scientific problems, have not yet adopted the dominant paradigm of their disciplines, and should therefore be more likely to cause scientific revolutions. This first group also provides support for Simonton's model of creativity (Simonton 1984, 1997), stating that each individual has an initial 'creative potential' that decreases over time.

In contrast, the second group of studies argues that it is not the younger researchers, but the mid-career and older researchers who produce the most research and have a greater scientific impact (Dennis 1956; Cole & Cole 1973; Allison & Steward 1974; Cole 1979; Wray 2003, 2004; Kyvik & Olsen 2008). This follows the Mertonian (Merton 1968, 1973) argument, according to which, as scientists rise in the science hierarchy (Cole & Cole 1973), they obtain more funding, supervise more graduate students, gain access to more resources and, hence, increase their productivity and impact.

Gingras et al. (2008) showed that both theories have some merit: Older Quebec professors are more productive (they are the leaders of their own teams and have access to more and better resources), but younger scientists have more impact (measured by citations).

Allison and Steward (1974) found that highly productive scientists remain productive as they age, but those who produce little, publish even less later on. The starting conditions of one's career therefore seem to matter. Lehman (1953) also highlighted the disciplinary differences in scientists' productivity peak. There is evidence that the effect of age varies between different disciplines (Pelz & Andrews 1966; Kyvik 1990; Levin & Stephan 1991).

Our main aim in this chapter was to test the relationship between age, gender and scientific field and publications, as reported by our respondents. Could our study present unequivocal evidence either way: Whether younger or older academics are more productive and if so, whether this is correlated with the gender and scientific field of the respondent?

Reported volumes of research publications

The results presented here refer to 'reported' publications in different categories. Although it is always possible that self-reporting of publications can introduce some degree of bias (over- and even underestimating numbers of publications), we believe that the size of our sample mitigates against any major bias at the aggregate level. We asked respondents about the volume of their research output over the three years prior to data collection. Respondents were asked to provide information about six categories of output (Table 25).

Table 25: **Median publication outputs by scientific field**

Field		Articles	Books	Book chapters	Conference papers	Policy documents	Popular articles
Natural sciences	N	1 286	818	879	1 023	743	916
	Median	6	0	0	2	0	2
Agricultural sciences	N	493	354	384	449	340	405
	Median	6	0	1	3	1	2
Engineering and applied technologies	N	524	341	354	477	299	367
	Median	4	0	0	4	0	1
Health sciences	N	959	754	797	874	786	810
	Median	7	0	0	2	1	1
Humanities	N	249	196	220	176	146	170
	Median	5	1	2	1	1	2
Social sciences	N	846	623	736	717	624	653
	Median	5	0	1	2	1	2
Total	**N**	**4 357**	**3 086**	**3 370**	**3 716**	**2 938**	**3 321**
	Median	**6**	**0**	**1**	**2**	**1**	**2**

The results in Table 25 – which disaggregated reported output only by scientific field – already present some interesting, and even counterintuitive, results. The medians of reported number of articles do not vary hugely by field (with the median for the health sciences highest, at 7). Respondents in the social sciences and human sciences were the only ones (as one would expect) that reported any significant numbers of published books and book chapters. Another expected result is the fact the respondents in the engineering sciences reported the highest number of published conference proceedings. Policy documents were listed by all respondents, except for those in the natural and engineering sciences. And finally, the fact that most respondents reported having written popular articles may be an indication that African academics also recognise the value of popularising their research results. In the remainder of this chapter we report in more detail on each of the main categories of research publication, and the results of three-way analyses of variance.

Scientific articles: Age, scientific field and gender

Our first set of analyses focused on the reported number of scientific articles in peer-reviewed journals. The main finding is in line with much of the scholarship indicating that, on average, younger respondents reported having produced a lower mean and median number of articles in the preceding three years (6.9 and 5) than their older counterparts (8.5 and 6) (Table 26). As article output tends to be influenced by field as well as gender, we controlled for these two variables. A three-way between-group analysis of variance was conducted to explore the relationship between age, scientific field and gender with number of articles in academic journals. The interaction effect between the three independent variables was found to be statistically significant, F (10, 4 184) = 1.968, p = 0.33. The results in Table 26 show that the main effects (age and output; gender and output; and field and output) were all significant at P<0.00. In addition, the largest interaction effect was found between age, field and output.

Table 26: **Reported article output, by age, field and gender**

		39 or younger			40–50			Older than 50		
Field	**Gender**	**Mean**	**Median**	**N**	**Mean**	**Median**	**N**	**Mean**	**Median**	**N**
Natural sciences	Male	7.0	5	303	8.9	6	359	10.9	7	250
	Female	5.5	4	136	7.3	5	115	7.3	6	69
	Total	6.5	5	439	8.5	6	474	10.1	7	319
Agricultural sciences	Male	6.7	5	107	9.6	8	146	9.2	8	115
	Female	4.6	4	30	5.8	4	63	8.9	7	23
	Total	6.2	5	137	8.4	6	209	9.1	7	138
Engineering and applied technology	Male	5.2	4	137	7.5	5	135	9.6	6	137
	Female	2.4	2	36	6.5	4	42	14.4	9	21
	Total	4.6	3	173	7.2	5	177	10.2	6	158
Health sciences	Male	7.6	6	166	11.1	9	255	9.3	7	155
	Female	6.0	5	130	8.4	7	129	8.3	7	102
	Total	6.9	5	296	10.2	8	384	8.9	7	257
Humanities	Male	7.4	8	25	7.1	5	48	5.8	5	70
	Female	4.2	3	20	4.1	4	31	5.9	4	41
	Total	6.0	4	45	6.0	5	79	5.8	5	111
Social sciences	Male	6.4	5	142	6.5	5	205	6.5	5	157
	Female	4.4	4	94	5.4	5	127	5.6	5	98
	Total	5.6	4	236	6.1	5	332	6.2	5	255
Total	Male	6.7	5	880	8.8	6	1 148	9.0	6	884
	Female	5.0	4	446	6.7	5	507	7.5	6	354
	Total	**6.2**	**5**	**1 326**	**8.2**	**6**	**1 655**	**8.6**	**6**	**1 238**

Older respondents, irrespective of scientific field or gender, reported having published higher numbers of articles in the preceding three years. Those who were older than 50 at the time of the survey, on average reported having published a median of 8.6 articles in the preceding three years, compared to a median of 8.2 articles for those between 40 and 50 and 6.8 for those 39 years or younger.

Figure 54: **Mean reported article output, by age, field and gender**

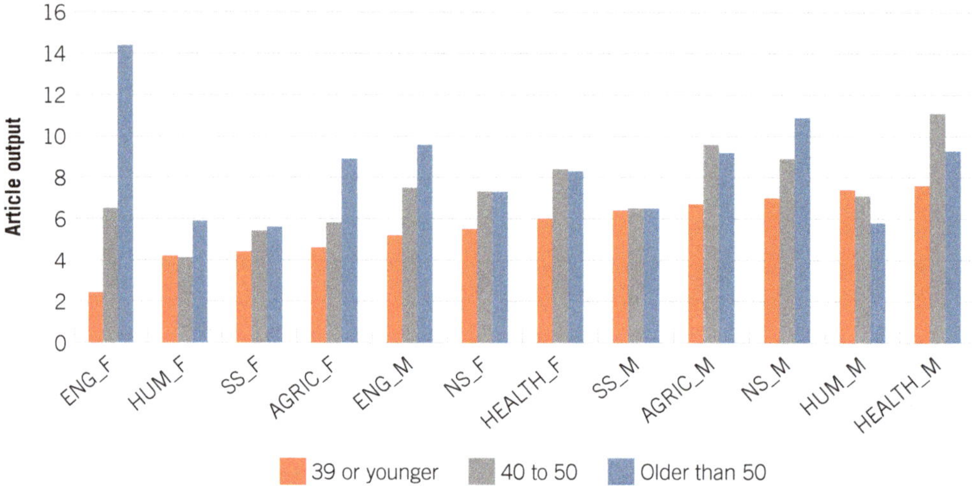

When controlling for field and gender, male respondents in general (with the exception of older respondents in the engineering sciences) reported that they had published higher numbers of articles.

Books: Age, scientific field and gender

The reported numbers of book publications, as disaggregated by field and by age interval, are as expected (Table 27). The averages for the humanities and social sciences are significantly higher (for all age intervals) than for the other fields. The interaction between age, gender and scientific field (Figure 55) presents more detail. Not only did respondents in the social sciences and humanities report higher numbers of book publications, but older respondents – in general – did the same. What is interesting is that the ANOVA results did not show any statistically significant differences between women and men academics (when controlling for field and age).

Table 27: **Reported book output, by age, field and gender**

		39 or younger			40–50			Older than 50		
Field	**Gender**	**Mean**	**Median**	**N**	**Mean**	**Median**	**N**	**Mean**	**Median**	**N**
Natural sciences	Male	0.4	0	175	0.6	0	232	0.8	0	176
	Female	0.2	0	76	0.4	0	67	0.5	0	52
	Total	0.4	0	251	0.5	0	299	0.8	0	228

		39 or younger			40–50			Older than 50		
Field	Gender	Mean	Median	N	Mean	Median	N	Mean	Median	N
Agricultural sciences	Male	0.5	0	67	0.8	0	111	1.6	1	92
	Female	0.3	0	18	0.3	0	39	1.3	1	23
	Total	0.4	0	85	0.7	0	150	1.5	1	115
Engineering and applied technology	Male	0.4	0	83	0.5	0	94	1.0	0	97
	Female	0.3	0	22	0.4	0	23	0.6	0	11
	Total	0.4	0	105	0.5	0	117	1.0	0	108
Health sciences	Male	0.3	0	122	0.5	0	206	0.8	0	137
	Female	0.2	0	93	0.4	0	103	0.5	0	78
	Total	0.2	0	215	0.5	0	309	0.7	0	215
Humanities	Male	1.5	1	21	1.2	1	38	1.4	1	67
	Female	0.5	0	11	0.5	0	17	0.8	0	32
	Total	1.1	1	32	1.0	1	55	1.2	1	99
Social sciences	Male	0.7	0	93	1.1	1	161	1.3	1	131
	Female	0.3	0	65	0.6	0	80	0.7	0	74
	Total	0.5	0	158	0.9	0	241	1.1	1	205
Total	Male	0.5	0	561	0.7	0	842	1.1	0	700
	Female	0.2	0	285	0.5	0	329	0.7	0	270
	Total	**0.4**	**0**	**846**	**0.6**	**0**	**1 171**	**1.0**	**0**	**970**

Figure 55: **Mean reported book output, by age, field and gender**

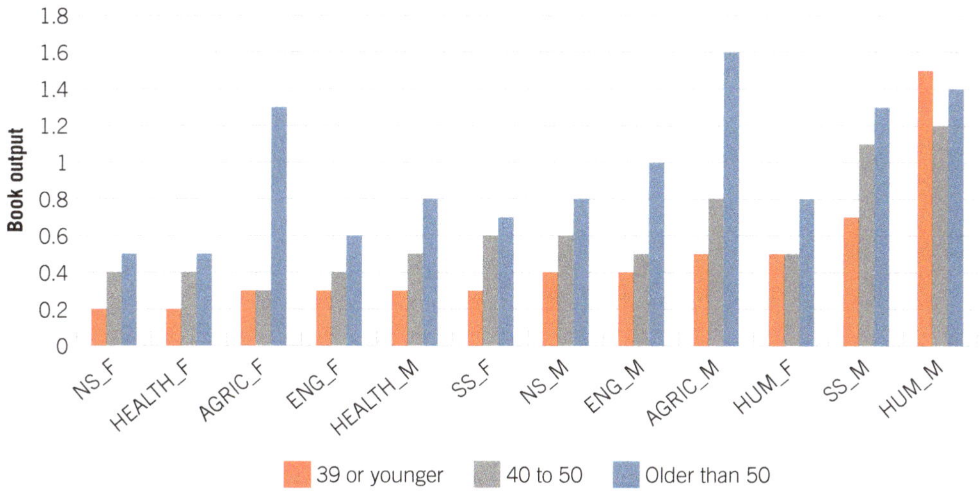

Conference papers: Age, scientific field and gender

It is a well-established fact that conference proceedings are a preferred publication outlet in certain disciplines, most notably those in fields such as mathematics, computer sciences and the engineering sciences. Our results conform to this conventional wisdom, with respondents in the engineering sciences (especially older male respondents) reporting much higher average numbers of conference proceedings than their counterparts in other disciplines. Their average reported output of 4.5 (for younger respondents), 5.5 (for those between 40 and 50) and 9.0 (for those above 50) is nearly double that reported by the respondents as a whole. The ANOVA results show a statistically significant interaction effect between age and field. Figure 56 clearly shows that older respondents are more likely to have published a higher number of conference proceedings. However, no statistically significant gender effect was found for any of the fields. Despite this, it is interesting that older women academics in engineering (compared to their male counterparts) reported a much higher number of conference proceedings. It should be added, however, that their number is quite small and may account for these 'anomalies'.

Table 28: **Reported conference-paper output, by age, field and gender**

		39 or younger			40–50			Older than 50		
Field	**Gender**	**Mean**	**Median**	**N**	**Mean**	**Median**	**N**	**Mean**	**Median**	**N**
Natural sciences	Male	3.1	2	225	3.7	2	306	4.8	2	204
	Female	3.2	2	93	3.1	2	89	5.3	3	62
	Total	3.1	2	318	3.5	2	395	4.9	3	266
Agricultural sciences	Male	3.0	2	87	4.1	3	139	5.0	3	110
	Female	2.0	1	27	3.8	2	56	6.0	5	23
	Total	2.8	2	114	4.0	3	195	5.2	4	133
Engineering and applied technology	Male	4.7	3	130	5.3	4	126	8.5	4	117
	Female	3.9	2	31	6.2	4	39	12.1	8	19
	Total	4.5	3	161	5.5	4	165	9.0	4	136
Health sciences	Male	2.9	2	151	4.2	3	239	4.8	3	153
	Female	3.4	1	108	3.0	2	118	3.4	2	89
	Total	3.1	2	259	3.8	2	357	4.3	3	242
Humanities	Male	2.1	1	20	3.2	2	36	2.2	1	54
	Female	1.1	0	11	1.9	1	18	2.2	1	31
	Total	1.7	1	31	2.8	2	54	2.2	1	85
Social sciences	Male	2.0	1	121	3.2	2	185	3.3	2	138
	Female	2.1	1	72	2.5	1	98	2.8	2	83
	Total	2.1	1	193	2.9	2	283	3.1	2	221
Total	Male	3.1	2	734	3.9	2	1 031	5.0	3	776
	Female	2.9	1	342	3.3	2	418	4.2	2	307
	Total	**3.1**	**2**	**1 076**	**3.7**	**2**	**1 449**	**4.7**	**3**	**1 083**

In summary, the publication behaviour of African scholars and scientists is, in general, in concordance with the conventional wisdom and scholarship in bibliometrics. The age and gender of the respondents, well as their scientific field, were shown to be the most significant predictors of reported output across all categories. More specifically, our survey results support the view that publication output increases with age, and this was found (in general) to be independent of field. As far as the gender of respondents is concerned, male respondents (with a few exceptions) reported higher numbers of outputs (irrespective of field and age). As far as specific types of output are concerned, the reported result for articles, books and book chapters, and conference proceedings were mostly in the expected direction.

Figure 56: **Mean reported conference-paper output, by age, field and gender**

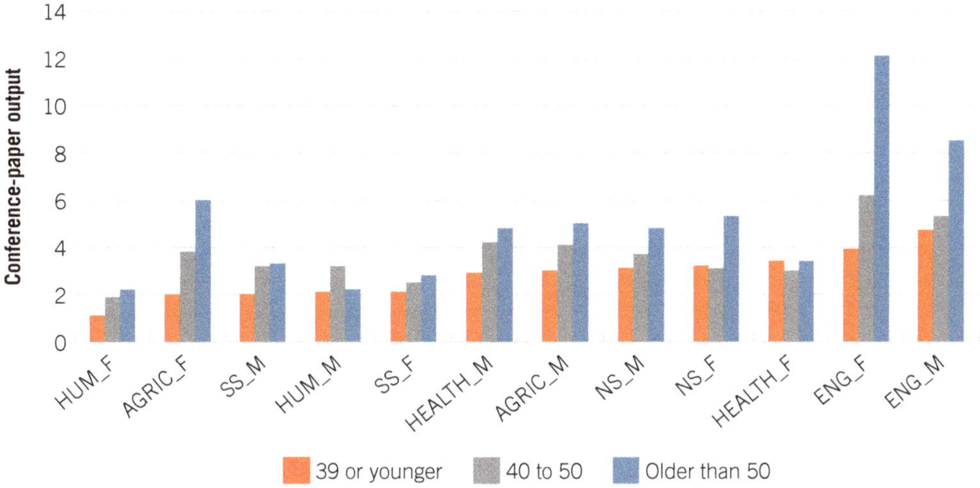

In the remainder of this chapter we present some of the results of the qualitative interviews as they pertain to issues around scientific publication.

Enablers of and barriers to scientific publishing

Four main themes emerged from the interviews with young respondents: (1) the prevalence of a research performance culture, and hence the pervasive demand for publication; (2) the consequence of the 'publish or perish' culture; (3) the barriers to publication as perceived by young scholars; and (4) the areas young academics identified as ones in which they need training and support.

The demand to publish

It is generally assumed that, at most universities, academic promotions are increasingly informed by research performance assessment. Appraisal of the research work and output of academics is increasingly valued, often at the expense of performance in teaching and other areas, such as university administration and service rendering. It is not surprising, given the increasingly widespread adoption of research performance policies and processes, that university incentive and reward systems remain increasingly biased in favour of research (and publication) performance, rather than rewarding other academic activities. It is clear from our interviews that young academics at African universities also experience this as an increasing pressure. A 38-year-old male respondent from Uganda stated the following:

> There's a saying that you either publish or perish. ... So, in academia, if you're going to remain active in academia, you have to publish. Otherwise if you don't publish then you're not going to be known internationally. And, of

> course, I think part of the perishing is that you won't have work to do. And if you have no work, then you won't have anything to eat as well.

For many interviewees the pressure to publish arises from the desire to attain more favourable working conditions or to enhance their employability.

> And maybe I am wrong. Is that [publishing in journal articles] gives you considerable power to negotiate? Because now you are bringing in a certain amount of income which the university doesn't want to miss out on. So, you are in a better position to negotiate around, wait a minute: my teaching load is too high. I only want to do this … So, I think this puts them in a better position to negotiate around their teaching. Around their supervision. And to create those spaces for themselves. Whereas those of us who are still kind of, you know, still struggling, it limits our ability to negotiate around those issues. (45-year-female from South Africa)

> I understand if I want to stay in academic areas and look for job somewhere else, publication is mandatory for me. So as a means or as a prerequisite to look for jobs somewhere else in a developed nation, I think that publication is a requirement for me. (36-year-old male from Ethiopia)

The pressure is also noticeably felt at institutions where publishing is a primary factor determining promotion.

> Incentive is always there because … you cannot be promoted to another rank of an academic career without a publication. So it's only publications that can make you go to the next level. For example, if you're a lecturer, you go to senior lecturer, you must have published at least a minimum of four papers, depending on how many members you are publishing together with. But at least with four papers, you can go to the next level. (39-year-old male from East Africa)

> So even if the university is not say anything, once you mention publications, they look at the name of the journal and the impact factor, and if it's something that is not good enough, then they are not going to give you a promotion, then they are going to play around. So, that is how it is. (34-year-old male from Uganda)

Lack of a conducive research environment

The historical function and focus of universities and research institutes impact directly on the research environment. Institutional policies and guidelines may be inadequate, and may even disincentivise local scientists from undertaking teaching and research. Research approval systems are cumbersome. Staff lack experience and expertise. Junior staff (especially from West Africa) seem to distrust senior staff. Some researchers are also distrustful of the motives of some journal editors' acceptance of journal articles for publication, and suspect them of bias.

Institution type

The nature and focus of an institution may also impact on the career and research prospects of scientists in Africa. Interviewees from Nigeria discuss the distinction between private and public research institutes, and specifically, that private institutes receive less state-sponsored support.

> For the public university there is funding by the government, by the Federal Government of Nigeria in the form of TED fund, Tertiary Education Fund, for short TED fund. And this fund [is] made available for some, for university scholars who are interested in carrying out one research or the other. So some universities make this fund available for the staff that are interested and some also siphon, some don't spend the money for the purpose. But this fund is not made available to the private universities. So that's a big challenge that the private university is really facing. So it's really affecting research in that area. (39-year-old from Nigeria)

> I am in a private institution … It's not a place you would want to be for a long time, because of the limited funding, and instrumentation, and all that. … Something I would want, to move to a better institution. Yes, a better institution has the opportunities. There are no vacancies, and … so you're kind of getting stuck somewhere, and just believing that opportunities will open up somewhere that you can move to, and have some financial motivation, have experienced people to work with. And have a better lab to work with, and all that. (30-year-old male from Nigeria)

In South Africa, historically disadvantaged institutions play a major role in researchers' ability to conduct research.

> I think the reason that I can say is only that the university itself has been a previously disadvantaged university. I did not want to go there that much. I think there was no promotion of research since a long time ago. Changing that particular culture over time is going to take some time. I think that's the main reason why research is not supported. (37-year-old male from South Africa)

Also in South Africa, the historical emphasis on technical training (and less focus on research) at universities of technology has resulted in staff currently not optimally capacitated to conduct research.

> I think the problem here is that we have a lot of novice researchers. … We don't even have a professor in our faculty. So, you know, there's a lack of research experience that can nurture younger people. In fact I think that's actually the

> biggest thing, that you don't have a pool of experienced researchers that can nurture and guide and facilitate a younger people coming into the structure. (40-year-old female from South Africa)

An interviewee expresses similar frustrations with regard to a newly established institution in Ethiopia.

> You know our university and most universities in my country are newly built and expanded, and they lack some experienced professors which mentor the newly coming staff in order to help teach how to write these winning proposals and these are very difficult. It's a problem. (38-year-old male from Ethiopia)

Lack of support for PhD and post-doctoral research

An interviewee from Zimbabwe relates his frustration in trying to support PhD and post-doctoral research.

> You see under normal circumstances in a nicely structured system it would not have too much of a bearing in the sense that if you've got a good structured system where you've got in place your post-doctoral fellows hopefully being funded publicly by the university and maybe through some other grants which are coming in, then your research can continue pushing on through your post-docs and PhD students ... But now the situation that we find ourselves in Zimbabwe which they've been trying to address is that that type of system is virtually non-existent and you have to build it by yourself without it necessarily being recognised by the university system. So I could have a post-doc and they come to work in my lab, but then the university structure and system as it stands now does not recognise that particular position officially. (40-year-old male from Zimbabwe)

Institutional policies, guidelines and incentives

Fixation with rankings

University ranking systems have reportedly led to a change in management culture, with a strong emphasis on quantifiable research outputs, which is out of touch with the daily realities with which lecturers and senior lecturers are faced.

> [My university] has particularly bad cases of ... managerialism that feels very out of touch and it's completely obsessed with rankings ... climbing university rankings all the time, and often out of touch with, yes, the realities of, first of all, the quality of the undergraduate students, the actual quality of them, and the kinds of strains that staff take and are taking ... Management really are kind of out of touch and they drive a lot of initiatives from above that often feel kind of, you know, pie in the sky and, and not even that often really seriously academically motivated, really. (36-year-old male from South Africa)

Promotion criteria

An interviewee from Cameroon feels that neither research nor teaching excellence determines progression and promotion.

> The factors are more because I say to myself, elsewhere perhaps excellence is recognised, if you are strong. Just be competent, it'll be fine. In Cameroon, beyond being competent and sometimes you do not even need competent to evolve. That's it, it's mostly the most negative aspect of our system, which is the most annoying is that you can do what you want, you can be brilliant, publish, do whatever you want. (Male from West Africa)

Research approval systems

Ethics committees and research offices

Interviewees underlined problems with ethics committees and research offices hindering research.

> That we have a research office that sometimes takes more than half a year for an industrial contract, to get that signed. I mean if, for me that's totally unacceptable for a university that claims to be research led, that if you manage to convince somebody in industry out there to invest here, that one of the big hurdles is that [my university] puts a signature under it. That's for me shocking. (34-year-old male from South Africa)

> I think yes, the policy environment needs to be relooked at. By policy environment I mean the levels of approval of some of these things. To me we are failing. We have too many approvals for researchers, and some of those guys don't even read those things, they just stamp them. Like we have something here we call, it used to give some money for funds because it's a national research fund. And they came up with a way to now control all researches in the country and their aim was [to] avoid duplication. But now in [unclear] you have to spend up to about US$60 to get an approval from the NRS for total rubbish. I think there needs to be less back and forth. Now we have the ethics committee at the university which has been for example mandated to be clear and give the clearance. Then after a week now you go for that letter to the national research fund, it is 400 kilometers, I told you, in Nairobi, why can't they capacitate this guy who has the time and the technical expertise, give him the certificate [unclear] on their behalf down here instead of having to travel for it to Nairobi. Especially for those of us who are not based in Nairobi. So small bits here and there we need a policy relook. (40-year-old male from Kenya)

The consequences of the publication culture

In order to meet the perceived demand to publish at all costs, African scholars are pursuing a wide variety of strategies to increase the number of their articles being published. One specific and increasingly prevalent strategy revolves around producing publications from a doctoral study. At many universities it has now become a requirement that a doctoral candidate must submit at least one or two journal papers for publication before a doctoral degree is awarded. This is especially true in South Africa, where publications are linked to the national reward system administered by the Department of Higher Education and Training. In ongoing research conducted at CREST, we have established that at least 40% of all doctoral dissertations result in scientific papers and that this trend is on the increase. The following two quotes from South African academics illustrate the point:

> The majority of the publications that I have were based on my postgraduate studies. So from honours – because I've also managed to publish my honours work – and then I have now recently published three post PhD. (33-year-old female)

> Plus, the university expects that, if you're going to graduate for PhD you have to have a publication from your research studies. So without a publication you cannot graduate, so that was another motivation. (27-year-old female)

The requirement to produce scientific papers from a PhD is not necessarily viewed in a negative light, but often as beneficial to their careers.

> Article route, yes … I like it that way because … you can have your articles published while you're busy with the PhD you know by the time the PhD is complete, you have your articles in process of publishing all you know, which is a very nice thing as well … But this way kind of forces you, writing in the article way. You choose your or you know you kind of earmark which journal you would like to publish it in and then you … You send it to them while you're busy with the PhD. (44-year-old female from South Africa)

> My PhD was also a PhD based on publication. So that, I think from that time, you know, and working in terms of the quality that was required in terms of publishing. That set me off in terms of doing scientific research to contribute to knowledge generation. (36-year-old male from Zimbabwe)

A common, and not unrelated strategy, is co-publication between supervisors and their students.

> Since I am a professor, I should be supervising postgraduate students. That is where I always adopt the PhD [by] publication approach then [we] have publication always coming from even the graduate students that I supervise. (36-year-old male from Zimbabwe)

The role of the supervisor is therefore, not surprisingly, considered to be central in encouraging and demonstrating how and where to publish.

> My supervisor; actually, all of it, because he's the one who was encouraging us to publish ... But a lot of help came from my supervisor who knew how to choose a journal in which one can publish in. And he also guided me on how to write, so I would write and give it to him and he would review, critique, bring it back and I would fix it and, yes, it wasn't all on my own. (27-year-old female from South Africa)

> I identified the article myself, but based on the motivation from our supervisors, because they also play a critical role, our supervisors, to always remind us that never allow your work or your research work to just be in delivery. Just try and, even if it's chapter one, literature review, try and see which may be [a] relevant article one can submit that. (44-year-old female from South Africa)

The demand to publish, especially from a doctoral dissertation, is clearly seen as contributing to building an academic career. However, there is also growing evidence that the emphasis on publication has negative consequences – some of which are quite unintended. One such negative consequence relates to the quality of scientific papers. Interviewees remark that, as the focus has shifted to publishing to attain promotion, there has been a decline in quality of output and a disincentive to undertake 'ambitious' research.

> One of the most dangerous things, so one of the biggest problems, at the moment universities see this issue of quality versus quantity. And I think there needs to be a way to, obviously, to try to balance those two. So in other words, when this kind of support is offered, there needs to be some expectation that researchers who are gaining from this should be able to produce some sort of output in terms of journal articles and so forth. And there needs to also be emphasis in terms of quality of papers that are coming out of there, because, unfortunately, at the moment in this country research that's funded, there is nothing that really ... there is no incentive for you to go the extra mile in terms of trying to be very ambitious, right? And hence right now we are sitting with the problem of predatory journals, and people are just under pressure to publish. (38-year-old male from South Africa)

> I've spoken with a few colleagues and oftentimes, you know, you're publishing ahead of the next promotion, which is a problem in itself, rather than kind of doing your work so that you can contribute a bit to making the world a better place, as it were. (40-year-old female from Nigeria)

Interviewees also expressed frustration with international journals and journals with high impact factors being prioritised over local journals. This negatively impacts on young scholars who are still learning how to publish. It also, often, leads to local journals being ignored.

> I think I would say that in terms of research and publications, there has been more emphasis on international journals. We have a local journal, but they hardly acknowledge it, yet it's difficult for people from the local institutions to publish in international journals. So, that could be a problem, because some of the things that are written by young and up-and-coming researchers can only be [published] by local journals. International journals have their own standards, and we have had problems with that. So, that is also something that blocks people. There is a hindrance to our development. So, we also need to link into collaborations that promote local publications. So, not only South African journals, but maybe East African journals, maybe a Ugandan journal. (34-year-old male from Uganda)

> Also, there's … this emphasis on the ISI journals [which] is troubling. … Some of my research is very particular, so a South African context, but I get no recognition if I publish in the South African journal … In fact, I've been told that I must not publish in that journal because it will look bad … So one of the journals that I, that would be a suitable avenue is not accredited, so I don't know what that process involves. There are others which are accredited, but they have low impact factors, and so I've been told it's better not to publish than to publish in a journal like that. (44-year-old female from South Africa)

One serious consequence of the pressure to publish has been a recent trend to increasingly publish in 'predatory' journals. Although there is now widespread awareness, and presumably also knowledge, of what predatory publishing is, it is still important to have a clear understanding of what is meant by predatory publishing and how it is defined. The term 'predatory publishing' is usually attributed to Jeffrey Beall – a librarian at the University of Colorado in Denver (USA). Beall, who has until recently been regarded as the unofficial 'watchdog' of predatory publishing, administered a website entitled 'Scholarly Open Access: Critical analysis of scholarly open-access publishing'. In his first major publication on the topic, Beall (2012: 179) provided a first description of what is meant by predatory publishing:

> Then came predatory publishers, which publish counterfeit journals to exploit the open-access model in which the author pays. These predatory publishers are dishonest and lack transparency. They aim to dupe researchers, especially those inexperienced in scholarly communication. They set up websites that closely resemble those of legitimate online publishers, and publish journals of questionable and downright low quality. Many purport to be headquartered in the United States, United Kingdom, Canada or Australia but really hail from Pakistan, India or Nigeria. Some predatory publishers spam researchers, soliciting manuscripts but failing to mention the required author fee.

Beall uses the term 'predatory' to refer to journals that 'prey' on (often unsuspecting and often young) scholars to submit their manuscripts for the sole purpose of making money from these scholars. In this process, normal good editorial and review processes are violated or suspended. Because these journals typically do not undertake any peer review (or very superficial) peer review, they are able to accept large numbers of manuscripts within very short turnaround times and hence make their money through high volume. Beall's point is that predatory journals and publishers are in the business of defrauding scientists and scholars.

In a recent paper, Mouton and Valentine (2017) showed how prevalent predatory publishing has become amongst South African academics. In their paper, they presented a more systematic comparison of the characteristics that distinguish 'normal' and 'standard' publishing from 'predatory' publishing.

Table 29: **Comparing the characteristics of good practice in scholarly publishing with those of predatory publishing**

Category	Standard publishing practice	Predatory publishing
Business model	Legitimate scholarly journals do not exist solely for profit	Predatory journals are open access journals that exist for the sole purpose of making a profit
Origin of papers	Authors usually submit manuscripts to journals of their own accord	Predatory journals typically solicit manuscripts by spamming researchers (especially using their Yahoo and Gmail accounts)
Journal titles	Legitimate journals usually have field- and discipline-appropriate titles	Predatory journals often have bizarrely broad titles (e.g. the *Global Journal of Advanced Research*) or titles with disjointed scopes (e.g. the *Journal of Economics and Engineering*)
Time to publication	Publication lag time is often correlated with the status of the journal (with the best journals taking more time to get to production because of high demand)	These journals boast extremely rapid (and unrealistic) response (review) and publication times. They often also publish extremely high numbers of papers per year. This is arguably one of the best indicators of whether a journal is predatory or not, as it speaks to the capacity of any editor to handle literally hundreds of submissions per year through proper peer review
Journal metrics	Journals indexed in Clarivate Web of Science and Elsevier Scopus have well-defined and transparent impact factor values	These journals boast extraordinary and often fake journal impact factors as well as false claims about where the journal is indexed
Peer review (stature of editorial board)	Legitimate journals have editorial boards and editorial procedures that properly oversee the process of peer review	Predatory journals very often have fake editorial boards or – at best – editorial boards that consist of a small number of individuals from the same organisation or country. They often enlist members of editorial boards who are not experts in the field. They also often include scholars on an editorial board without their knowledge or permission
Contact information	Legitimate journals provide accurate and appropriate contact information about their journal and editorial board	Predatory journals often list false or insufficient contact information, including contact information that does not clearly state the headquarter's location or misrepresents the headquarter's location (e.g. through the use of addresses that are actually mail drops)

The results of the Mouton and Valentine study showed that about 4% of all papers published by South African academics between 2005 and 2014 appeared in predatory journals. Their study also showed that over time the prevalence of publication in these journals increased.

During the interviews with young African scientists and scholars the issue of predatory publishing was mentioned repeatedly. A 34-year-old male academic from Uganda argued that African scientists are tempted to publish in these journals as they are unable to publish elsewhere.

> English is our second language. So, just getting accepted is difficult, and that's why you see these fake [journals] that are buying articles and publishing them, they are fakes, because they know people are desperate to publish to get their work accepted. So that is another problem, that the journals that are recognisable, it's extremely difficult. It could take you more than 18 months to have your article accepted. There are so many reasons why it's not getting accepted. It's extremely common that the stuff you are writing, they have their own rules, so that is one of the things that I see as a barrier, so someone just gives up.

Not surprisingly, the main reason why academics publish in predatory journals is related to the pressure to 'not fall behind':

> Over the last five years, this issue of predatory journals has really come to the fore. And if I look just within my school, there has been one, two, in very recent history, within the last six months to one year, there has been three colleagues who have been promoted. All of whom have published in predatory journals, and where the predatory journal publications that they have published in were not discarded when they were considered for promotion. ... The thing with publications with predatory journals is that it's easy, and you can get lots of them. So people who chose not to do that have been left on the back foot. So this idea that you need to publish as much as you can, as quickly as you can, people have sometimes bought into that, and that has actually served them very well ... So produce lots of articles, it doesn't really matter if they're in horrible journals, but that will serve you well. Whereas people who produce less articles, or publish it in lower amounts, and better-quality journals, that hasn't really been considered. (38-year-old female from South Africa)

Another interviewee cited the limited funding that is available for publishing in standard, high-impact journals as another reason why people publish in predatory journals.

> Yes, it is I can say, because it is not very adequate, considering that you can use it for other purposes. Like now, if I want to publish with a good impact factor, they may ask me maybe over US$1 000 and I can't pay that. US$1 000 is more than my two months' salary. (35-year-old male from Nigeria)

As elsewhere in the world, there is now widespread and growing awareness of the phenomenon of predatory publishing. Efforts are increasingly being made to help especially young researchers distinguishing between legitimate and 'fake' journals. These include the posting of notices, the provision of training and the implementation of institutional checks and balances.

> In the last few years the focus has shifted more to doing quality research and publishing with reputable journals or publishers and there was brought in some more checks and balances ... to avoid predatory journals or publishers ... If someone, even if it is out of ignorance, publishes with a predatory journal, that research does not get recognition or anything like that, and for instance, if you, if a post-graduate student publishes from their master's dissertation or PhD thesis the supervisor of the study should sign off and they have to consult with the supervisor in terms of where they would publish, whether it's a reputable journal or if it's with a publisher, if it's an academic publisher. ... So, there're people who have to sign off on, especially for emerging researchers who are not established yet and shortly after post-graduate studies to help these people to make responsible decisions in terms of where to publish and how to go about that. (29-year-old female from South Africa)

> I've attended a session here at the university, recently on predatory publishing ... But apart from that, when this entire thing started, we got regular updates from the institution explaining what predatory publishing is and how to identify these journals. (Male from South Africa)

The barriers to publication

Young academics often do not understand how the scientific publishing system works. Unless they receive some guidance or support in this regard, it is often the case that they struggle to get their first manuscripts accepted. In addition, being inexperienced in scientific publishing also typically leads to unrealistic expectations. One example of this we found in our study, is the fact that a number of young scientists expressed frustration at the long delays in getting their articles published.

> And the final thing, which I guess is outside of anyone's control, is the actual time to publish. So, I mean, yes, that basically is based on the journal. You might get some journals that respond in two weeks, or two years. You know, that's a slightly different out-of-scope idea. But, that being said, you can at least say, here is a pipeline. I have three working papers, and they are under review, or whatever the case might be. (28-year-old male from South Africa)

> So I am aware that it might take some time, but I mean, since May – it's November now, so! ... You know, I don't know how long should I wait, but ... And the worst thing is that the status it keeps on showing me that it's pending,

> so I don't know. I mean, there's no communication since May. I don't know, I don't know. (27-year-old male from South Africa)

Some interviewees expressed their distrust of the editorial process. Editorial ignorance of the African context is suggested to be a reason that African publications do not receive due regard and consequently African academics are becoming discouraged.

> I will mention the peer-review process. I do my research in Africa, in particular the African context, and I send to a[n international] journal … and then they send the paper to a reviewer in the US or in Holland or somewhere else. They don't really understand the African context in terms of the issues, in terms of some of the challenges that we face, like, on the ground, and some of the comments that you get from the review process, they're very depressing if I can use that word. So for a researcher to take those comments, it's very depressing and you kind of lose your motivation, because you are on the ground, you see those issues and you want to report or write about them, and you really feel very strongly that this is a problem in Africa, but probably it's not a problem in Canada, it's not a problem somewhere in Europe. So it kills the drive in many researchers and it also explains probably why many African researchers end up in those pirate journals now, where the peer-review process is not that thorough. So I think the attitude and mindset of some of the researchers, from my personal experience, is actually negatively affected my research projects. And I'm sure even some of my colleagues, they have this sentiment, they say if you send to this international journal, they send to reviewers who don't understand what is happening in Africa, they always reject, so the journals, they're like for Europeans or Americans, a certain group of people, which then also affects the quality of our research, particularly in terms of publication and communication. (35-year-old male from Zimbabwe)

Uncertainty regarding the reasons for article rejection causes doubt and even suspicion.

> I'm doing the articles route for my PhD and its publishing in journals. I know it's a normal experience but when you submit your manuscript to journals it will rejected and so on, but that uncertainty of, you know, what is it that the journal or the editor actually wants, I find that that lack of objectivity is worrying because it's supposed to be an objective process but I find that in the end the editor makes a decision even if there is a peer-review process. And I've had instances twice now with my four articles where there seems to be a difference of opinion between reviewers and editor and it's like okay this is what the editor says and there isn't … The comments suggested there's nothing wrong with my actual content, the actual study meets all the ethical and process requirements but as I said the editor wants it done in a certain way. For example, with my third article the editor actually said put these words, these five in your conclusion, start your sentences with these words.

> And I was like but I only did that when I was teaching at primary school. Sentences. I'm doing a PhD and this person is telling me how to start my sentences. But he doesn't say that there's anything wrong with the content. (48-year-old female from South Africa)

The result of consistent rejection has translated in discouragement for two interviewees.

> I must say, in terms of my PhD research, I haven't yet published anything out of it because I've had a lot of rejections … I write the article, I send it to them, they make a few corrections, they say yes, okay, it's fit for publication, or it's fit for journal, it goes to the journal, I do the corrections, it comes back, and then they say oh no, it doesn't, you know, we're not interested in it or whatever. So, I think that they, you know, are wasting my time and I've got other more important things to do, students that are in my face and stuff. So, I'm feeling very negative about this whole publication story. (South African respondent)

> It's also very hard to publish. I mean let's face it. In South Africa, you have to understand the, kind of, rules around publishing. You have to understand a lot about the journal. About how the students fit in. I'm ready to submit this to the journal. I said, well, did you read what the journal mission statement is? Did you see what is it that they are taking? … So, it is always a concern about how do you, you know, when you are writing, is this going to be good enough to go in there? Have I understood what it is that this journal publishes? … And then you do wonder, you know, what chance do I have? And you put in months of effort. Lots of your time, your energy, into this. And to get the rejection, you know? … But how do I move forward? That's, you know, the big challenge, I think. (45-year-old female from South Africa)

The areas identified by young academics where they need training and support

Young scientists need to be taught the basics of journal publication. Which journals to publish in and what journals want from a publication are central concerns.

> I definitely think emerging researchers need help. For example, when I started I had no clue as to which journals to publish in and the ranking of the journals and I had no idea. I didn't know how to write a paper, nothing. So, I definitely think … it doesn't have to be a week-long just really a day or two days and you know journal writing and where do you find journals. How do you publish in these journals and what to look for etcetera, I think that would really help. (38-year-old female from South Africa)

The potential value of input from journal editors in this regard is emphasised.

> I think in terms of perhaps that sometimes these, the editors of these journals maybe do a road trip and go around and talk to us in, you know, so some of the more popular health science journals in South Africa, the local ones, I'm not even going international, but perhaps if those editors came around and gave talks about it, because, you know, I submit to my supervisor, say for my PhD, for example, my supervisors will look and they'll say oh yes, they think this journal will be good, it seems to, you know, the crux of the journal is appropriate ... You send your stuff, they send it back saying no, sorry, it doesn't match our thing. Or my recent experience was where I submitted the article, it went to reviewers, it came back with corrections, I did the corrections, I resubmitted it, and then they come back months later and say oh, actually, no, we decided that there's nothing new in this article, so we're not going to publish it. Why didn't you tell me that in the first place? So, I think perhaps if they come around and kind of talk to us a bit about, you know, some of the tips on publishing in their particular journal. (Female from South Africa)

> If these editors from various journals that are within the university can actually develop programmes and say, guys, we have got a research seminar focusing on the barriers to publication, or, guys, we have got a seminar that is looking at interdisciplinary research, you know, those kind of elements, if that can actually be created, then it creates a very conducive research environment. Otherwise, people just publish because the university compels people to publish. (43-year-old female from South Africa)

An example of input from journal editors at a South African institution is given.

> It was a faculty initiative, they blocked off days, you know where they said okay, all these people are not supposed to go to class, they must come and write. They must have a writing session and what happen was the journal, the editors of the journal came to give us feedback and we were all expected to be there. So, we made the corrections and everything on that day, on those manuscripts, everything and got it published. So, it was very hands on. (44-year-old female from South Africa)

Young scientists also need support to build up their writing confidence and to develop various 'writing styles' with appropriate language use.

> I suppose I think the number one thing is, for me in my experience, is that people are not prepared, academics are not prepared to actually be academic writers and this I mean is not about like the acts of sitting down and writing a paper, I think that's part of it but I think that a lot of them actually just really struggle with the confidence of writing. (42-year-old male from South Africa)

> It's probably learning the writing style because I've written lots of documents in practice while I was there, big reports for forensics and it's got a certain

> style. Then you come and you write learning material and you write exam papers. Totally different style again, okay. Move over to research, different style of writing, totally ... I suffered as if I had to go through a personality change to be able to do that writing. Okay so, I've mastered that writing kind of but if you look at the articles, it's different again from how a masters is written. It's at a different type of level and the English I must say is, okay, yes at an entirely different level again. So, probably to get it up to standard in terms of the writing style and the English because the content is there. I'm sure the contribution is there, it's just now to get it across on paper, so that someone will accept it. (42-year-old male from South Africa)

Young scientists need to know where they can get access to funding for research projects and then how and where they can publish.

> And then the other aspect, it's just like very practical things. Like how to write a grant. How to even, where to go, what websites are good to go to if you want to track what's good for you. What's relevant for you. (34-year-old female from South Africa)

> So you need to know something about where to get the information on what to publish, or how to access funds. Where does one learn that? You have to also know who to ask, but this is, I don't know. I learn this on my own. (29-year-old male from Ethiopia)

Required training may occur 'on the job'.

> Well, I can attribute that to maybe the place I work. It's in my workplace. ... So in my workplace, we have experts in the field of science, technology innovation, entrepreneurship, management. So we'd work in teams. So it makes it a bit easier for us to do it together. I didn't really learn it anywhere when I started working here, and it's mainly through ... discussing with your peers, maybe amongst yourselves, sort of. So no formal training. (38-year-old male from Nigeria)

A conducive research environment

Support at the highest level of university management has a major impact on promoting a conducive research environment.

> Let me say, in the past five years, there was somebody who was DVC of research who was a little bit more concerned about the research, who really, really wanted to take this university forward when it comes to research. He came and implemented quite a number of things. And then unfortunately, because of issues within the institution, that person left the institution, and then after the person left the institution, that is when ... now this started again, going down. (37-year-old male from South Africa)

It is noted that heads of department can play a pivotal role in research funding and staff motivation.

> One of our previous directors ... his main aim was to get research funding. And that was his main thing to do for the last 20 years. And based on his income that he generated for the department he spread all the researchers whether they obtained their own funding or not. ... Therefore all the researchers always had access to research funding. This created a good progress situation for all those who were capable of doing research. And I think that support that comes from a division or a department made a big difference and nurtured quite a few people, I mean to reach their ... how do I say, to reach certain levels in their career. (49-year-old male from South Africa)

> When I speak to my head of department, every time he motivates me, he always sees a star in me. So that is why I'm still, you know, pushing. It's actually my head of department who is motivating me ... and he is making sure that he supports me in anything that you can imagine, so that I can grow as a researcher. (35-year-old male from South Africa)

Conclusion

Our bibliometric analysis of research output in African countries (Chapter 2) over the past decade has shown that African scientists and academics have increased their output in international journals significantly over this period. The number of scientific articles has increased as has Africa's share of world scientific output. In this chapter we analysed the responses of young scientists as far as their reported publications are concerned. A more granular picture of the publication practices of African academics emerged from our presentation of the survey and interview findings in this chapter. In particular we see evidence of the challenges that young scientists experience in 'breaking into' the scientific publication system. We hear about the barriers to publication and the dangers of predatory publishing. We also see that many young scientists complain about a lack of a conducive and supportive research environment at their institutions. These experiences link to other institutional constraints that we have already discussed in the book. These include heavy teaching workloads, a lack of mentoring and training support, and a lack of mobility opportunities. We return to discuss these factors in our concluding section.

CHAPTER 9

Collaboration

Johann Mouton, Heidi Prozesky and Agnes Lutomiah

Introduction

Research collaboration is a sociological phenomenon which has attracted the attention of researchers and governments both locally and internationally (Yeung et al. 2005, cited in Pouris & Ho 2014). According to Wagner et al. (2002), governments use collaboration as a policy instrument for technology transfer between the universities and industry, for scientific and technological transfer from a foreign country, for enhancement of diplomatic relations with other nations and for political gains. Evidently, collaboration is argued to be a key element of science, technology and innovation policy, hence, governments support it through large investments (Wagner 2002; Pouris & Ho 2014). For instance, in the mid-1990s, the US government was estimated to have spent about US$3.3 billion on international collaboration. Particularly, the US government is estimated to have spent an average of US$322 million between 1994 and 1999 on collaboration with Russia, an amount that peaked in1996 at US$380 million and later decreased to about US$275 million in 1999 (Wagner 2002: 11).

Generally, the key factor behind the significance of scientific collaboration rests in its channelling of knowledge flows amongst scientists. Research collaboration has a central role in knowledge creation and innovation. Innovation and creativity are reliant on the presence of ideas which can create new knowledge, and collaboration is a key platform to harness and develop these important ideas (Katz & Martin 1997; Lee & Bozeman 2005). Toivanen and Ponomariov (2011: 473) argue that, 'this dynamic is particularly important for developing countries, such as many in Africa, with limited national knowledge stocks, infrastructure/instrumentation, and human capital'. In this case, collaborative research offers important channels for building up local scientific capacity (Katz & Martin 1997; Lee & Bozeman 2005). Collaborating both internationally and nationally with renowned scientists is claimed to be a great determinant for enhancement of scientific quality (Narin et al. 1991) and scientific output (see Borghei et al. 2013). Collaborative work is also claimed to result in faster diffusion of scientific knowledge (Ponds 2009).

In his review, Beaver (2001) listed increased synergy, feedback, dissemination, recognition and visibility as advantages of research collaboration. According to Beaver, collaborative research enhances feedback, recognition, dissemination and visibility amongst scientists. This is based on the assumption that each actor in the collaborative activity comes with a 'network' of fellow scientists who are keen on the research; each actor of the collaboration invested in the collaborative research is a visible member of the team; and that each individual comes with 'favorable reputation' to the collaborative research.

Scientific collaboration also enhances the credibility of research results as several scientists engage on the project. Furthermore, it is argued, collaborative work might 'reduce competition, increase trust, facilitate the exchange of complex knowledge, support the adaptation of a piece of knowledge, and help to speed up knowledge creation and innovation' (Gazni & Thelwall 2014: 261). Collaboration may enable knowledge exchange, transfer, use and sharing on the basis of the scientists' needs, goals, language, activities and understanding through their interactions (Gazni & Thelwall 2014).

Studies have also identified various advantages of collaborative research between researchers and practitioners. The advantages include facilitating access to data and the process of collecting data; researchers and practitioners become more familiar with each other's environment; improvements in the skills, practices and competency of the practitioners and researchers; practitioners identify with the researchers' viewpoints; research findings are put into use; and practitioners ensure availability of research grants (Jean-Louis et al. 2003; Ross et al. 2003).

Collaboration happens at various levels of the research system 'between research groups within a department, between departments within the same institution, between institutions, between sectors, and between geographical regions and countries' (Smith & Katz 2000: 33). Importantly, collaboration mainly occurs between individuals. Thus, the basic unit of research collaboration is deemed to be between two or more scientists. Basically, it is the people who participate in collaborative activities and not institutions (Smith & Katz, 2000). Inasmuch as the interpersonal collaborations are considered important, given that it is the people who collaborate at the several levels, Smith and Katz (2000: 33) observe that many of the policies aim to foster collaboration at the 'higher levels rather than inter-individual collaboration'.

Factors that influence research collaboration

Collaboration is influenced by several factors. This section discusses some of the main factors that influence collaboration. These factors include personal, scientific and technical factors among others. Personal attributes may include the demographic characteristics of collaborators which may either be a contribution or a hindrance to the collaboration process. These characteristics include age, gender and nationality among others (Bozeman et al. 2013). The assumption is that researchers who have the same demographic characteristics are more likely to collaborate with each other. Bozeman and colleagues have conducted a number of studies investigating personal attributes specifically gender as related to collaboration patterns (Bozeman et al. 2001, 2013).

Gender and collaboration

Gender is seen as one of the 'most personal and salient issues in one's life', particularly in academic science where there is under-representation of women and minorities (Pollak & Niemann 1998; Johnson & Bozeman 2012 quoted in Bozeman et al. 2013). Gender is said to be a key personal collaborator attribute in science. Bozeman et al. (2013: 8) note that, inasmuch as the influencing factors of female collaboration can be defined as the scientists' career attributes, 'the outcome of female collaboration is highly personal'.

A number of earlier studies in the literature, have shown that women scientists tend to collaborate differently and less effectively in comparison to male scientists (Cole & Zuckerman 1984). Studies have shown that female scientists are likely to collaborate less than their male colleagues (Sonnert & Holton 1996). Women are more likely to establish more formal collaborations (Sonnert & Holton, 1996), however the study by Bozeman and Corley (2004) showed that these collaborations and research networks tend to be less 'cosmopolitan'. Examining data from 451 scientists and engineers at academic centres in the US, Bozeman and Corley (2004) developed various models to understand collaboration patterns among academic scientists. One of these models examined the effects of gender, scientific field and tenure on the proportion of the female collaborators. Bozeman and Corley established that female scientists who are non-tenured, tenured, hold the rank of research faculty or research group leaders collaborate with a higher proportion (36%) of other females compared to the proportion (24%) of the male scientists in the same ranks. The analysis also showed that an overwhelming majority (83.3%) of 'non-tenure track females collaborate [more] with other females' (Bozeman & Corley 2004: 607). Another study by Rijnsoever and Hessels (2011) examined the characteristics of scientists that are linked with disciplinary and interdisciplinary research collaborations. They found that there seem to be changes in relation to gender and the collaboration patterns. Their results showed that women are more likely to be involved in interdisciplinary collaborations than men.

Bozeman and Gaughan (2011) conducted research to determine whether the observations discussed above on the differences in male and female collaboration patterns are linked to the 'actual differences' in gender or to 'false' relations associated with poorly developed models. The study used questionnaire data from a sample of US academic scientists (N=1 714). The authors note that, for instance, for most samples of academic scientists, female researchers tend to be younger than the males; therefore, the models need to account for this aspect to avoid distorted results. Given a dataset of 1 714 respondents weighted by gender and scientific field, Bozeman and Gaughan focused their analysis on research collaborations with industry and the motivations for collaboration. The study established, inter alia, that there are considerable gender differences in relation to the choice of strategies for collaboration. Men are more likely to lean on 'collaborations based on instrumentality and previous experiences' compared to females. The analysis also showed that having clear strategies for collaboration was connected with having more collaborators. Importantly, the study by Bozeman and Gaughan (2011: 1 393) found that 'women tend to have rather more collaborators on average' compared to men, especially when controlling for age, scientific field, tenure, doctoral cohort and family status and size.

Araújo et al. (2017), examining a dataset of more than 270 000 cases in Brazil, found large differences in gender in research collaborations. They established that, across all the fields analysed, male scientists collaborate more with other male scientists, whereas the females are more 'egalitarian'. This is in spite of the scientist's number of collaborators. The only exceptions were found in the field of engineering where, with an increase in the number of collaborators, the 'gender bias' disappeared (Araújo et al. 2017: 1).

It is clear from this review that the literature shows some conflicting findings on gender differences in collaboration (Bozeman & Corley 2004; Bozeman & Gaughan 2011). However, when other factors that influence scientific collaboration are controlled for, studies generally show that female scientists register a greater propensity to engage in collaborative and interdisciplinary research, they may have fewer collaborators and they tend to be less involved in international collaboration than their male peers (Cole & Zuckerman 1984; Sonnert & Holton 1996; Bozeman & Corley 2004; Van Rijnsoever & Hessels 2011).

Tenure and collaboration

Several studies have investigated tenure in relation to research collaboration. Tenure is deemed a key aspect of academic reward in academic institutions. The discourse on research collaboration always considers the need for one or more of the collaborators to have tenured positions (Boardman & Ponomariov 2007). Despite these debates, a number of studies on research collaboration revealed that tenure does not have significant effects on the collaboration choices or the number of collaborators. In their investigation, Bozeman and Corley (2004) established that tenure was not strongly and statistically significant related to the number of collaborators or the proportion of the female collaborators. Similarly, in that study, tenure does not seem to have a significant influence on the collaboration strategies. In addition, Bozeman and Corley (2004) found that tenure status was not statistically significant associated with the 'proximity' of researchers. When analysing the relationship between tenure and the collaboration choices and strategies, the authors observed that those who are untenured are more 'tactical' in their collaboration choices and strategies. The authors did, however, establish a statistically significant and positive relationship between tenure and the 'mentor' collaboration strategy (Bozeman & Corley 2004).

Further, Bozeman and Corley (2004) investigated the factors that determine the collaboration strategies for individual researchers. They established that 'the proportion of female collaborators, tenure status, the number of graduate student collaborators and the researchers' cosmopolitan scale' all have a statistically significant and positive relation with 'mentor' collaboration strategy (Bozeman & Corley: 607). That is, in terms of collaborating with graduate students, tenure track female faculty and tenured male faculty more often tend to participate in collaborative activities with graduate students.

Age and collaboration

Age is undoubtedly one of the personal factors that is likely to have an influence on research collaboration. However, there are few studies that have analysed the influence of chronological age and career age on collaborations. The assumption is that the influence of age on collaboration is 'obvious', that is, the older scientists are, the more they are likely to 'have more collaborators and a richer and more diverse collaboration network' (Bozeman et al. 2013: 7).

In relation to the aspects linked to collaboration, Rijnsoever and Hessels (2011) conducted a study with academic faculty at a university in the Netherlands. In their analysis, they established that research experience has a positive relationship with disciplinary and interdisciplinary collaboration. Apart from the personal attributes discussed above, scientific and technical factors were also found to impact on research collaboration.

The scientific field or disciplinary factors

Collaboration is influenced by disciplinary factors outlined by the nature of the work in a scientific field, as well as by the different traditions, cultures and practices of a given discipline (Melin 2000; Lee & Bozeman 2005; Fry 2007). Several studies have revealed that collaboration levels and co-authorships vary across scientific fields or disciplines (Katz & Martin 1997; Duque et al. 2005). Furthermore, co-authorship practices in different scientific fields are guided by different social norms. Melin (2000) notes that the readiness and need to collaborate, as well as the forms under which collaboration is done, varies between different scientific fields. For instance, in the medical sciences, scientists always work together in teams and often collaborate with other teams, while in the humanities, there are no teams and collaborations are uncommon.

In a study of 443 academic scientists at university research centres in the US, Lee and Bozeman (2005) investigated the factors that influence collaboration and sequentially examined how each impacts measures of productivity. From their analysis, there is evidence that the human capital attributes of individual scientists impacts collaboration; however, they found mixed results in relation to the influence of attributes on research productivity. Importantly, Lee and Bozeman established that scientific field has a significant impact on research collaboration. The study controlled for field differences as they classified scientists in two groups, 'basic' or 'applied'. 'Basic' fields comprised of physics, chemistry and biology while the applied fields consisted of engineering. In their analysis, the authors established a significant relationship between type of field and research collaboration, suggesting that engineering scientists collaborate more.

Given the above observations, other studies have also claimed that scientists in theoretical fields collaborate less and have lower productivity levels compared to those in 'experimentally-intensive' or 'applied fields' (Katz & Martin 1997; Lee & Bozeman 2005). In a study of US scientists, Lee and Bozeman (2005) observed high collaboration in the engineering fields compared to biology and life sciences. Also, Lee and Bozeman established differences in the productivity levels of the different fields with chemistry reporting the highest number of research publications and computer science the lowest.

From the body of literature reviewed above, the studies show that in fields like medical and natural sciences, scientists are more likely to work in teams and collaborate. The situation is different in the humanities and social sciences where scientists tend to work individually and collaborate less. Also, in fields such as high energy physics that exhibit greater levels of mutual dependence for knowledge, resources and skills, or have low degrees of task uncertainty and, where the task outcomes are clear, scientists tend to collaborate more so as to make significant scientific contributions. Such fields include high energy physics.

Reported collaboration by young scientists

In our study we asked respondents how often they collaborate with researchers in their own institution (intra-institutional collaboration), in other institutions in their country (national collaboration), in other countries in Africa (African collaboration) and outside Africa (international collaboration). They could rate the frequency of their collaborations with these different types of researchers on a five-point Likert scale (1: 'never or rarely'; 2: 'rarely'; 3: 'sometimes'; 4: 'often'; 5: 'very often'). For each category, we divided the respondents in two groups: those who collaborate 'often' or 'very often' with each type of collaborators, and the rest. In this chapter we focus mainly on the responses of the young scientists (39 years or younger).

The results (Figure 57) show that collaboration with academics and scientists within their own institution is by far the most frequent form of collaboration (62.9%), followed by near equal proportions that listed international collaboration (36.9%) and national collaboration (35.7%) as their preferred mode of collaboration. One interesting similarity between the survey results and the results of our bibliometric analysis of research collaboration discussed in Chapter 2, is the low priority given in both cases to collaboration with academics and researchers within countries on the African continent.

Figure 57: **Type of collaboration**

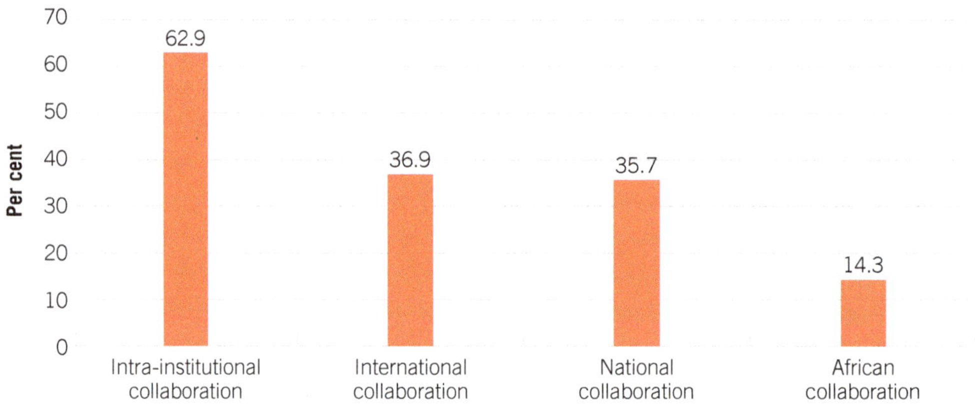

In the following section we test whether there is a relationship between collaboration type (as reported) and four variables: age, gender, rank and scientific field. The cross-tabulation between age and collaboration type (Table 30) showed very small (and statistically insignificant) differences.

Table 30: **Frequency of reported collaboration, by age category**

		39 or younger	40–50	Older than 50
Intra-institutional collaboration	Less than often	38%	38%	38%
	Often/very often	62%	62%	62%
International collaboration	Less than often	65%	63%	61%
	Often/very often	35%	37%	39%
National collaboration	Less than often	65%	64%	65%
	Often/very often	35%	36%	35%
African collaboration	Less than often	88%	84%	86%
	Often/very often	12%	16%	14%

As far as gender of the respondents are concerned, some statistically significant differences ($p<0,05$) were found, although the differences are not large. But the overall pattern is the same, with slightly higher proportions of young male respondents reporting higher frequencies of collaborations of all four types.

An analysis of rank by collaboration type confirms the findings of previous studies, as significantly higher proportions of staff in the senior ranks (especially the professoriate) reported more frequent collaborations of all types.

Table 31: **Frequency of reported collaboration by gender (often/very often responses)**

	Male	Female	Total
Intra-institutional collaboration	63.5%	58.9%	62.1%
International collaboration	37.9%	34.8%	37.0%
National collaboration	36.7%	32.8%	35.6%
African collaboration	16.2%	9.9%	14.3%

Table 32: **Frequency of reported collaboration (often or very often) by academic rank**

	Professor, associate professor		Senior lecturer		Lecturer		Researcher/ scientist	
	Count	Row N %	Count	Row N %	Count	Row N %	Count	Row N %
Intra-institutional collaboration	1 008	32.7%	578	18.8%	507	16.5%	613	19.9%
National collaboration	550	31.5%	298	17.1%	276	15.8%	385	22.1%
African collaboration	211	30.8%	129	18.8%	102	14.9%	163	23.8%
International collaboration	656	36.2%	305	16.8%	265	14.6%	372	20.5%

Our final analysis considered the relationship between being mobile (see Chapter 6) and frequency of collaboration. As one may expect, a strong and statistically significant relationship was found. Those who are more mobile (having studied or worked abroad) are also more likely to report higher intensity of collaboration.

Figure 58: **Relationship between mobility and collaborating internationally**

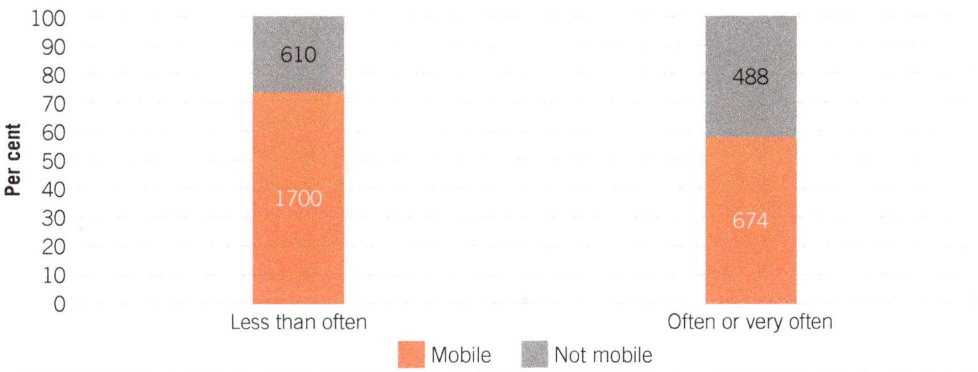

Comparison between collaborators and non-collaborators

There are no age differences, in general, between those who collaborate often or very often with each type of collaborator, and those who do not. However, the proportion of women is slightly higher among those respondents who collaborate less or not at all with researchers in Africa, while the gender difference is not significant for collaborations outside the continent.

Respondents who collaborate more are much more successful at raising funds, and much larger amounts, regardless of the category of researchers they collaborate with. The number of publications is always higher for researchers who collaborate more (the difference for books is not significant for researchers who collaborate more within their own institution or outside Africa).

Unsurprisingly, the more often a respondent collaborates with one category of researcher, the more prone he/she will be to collaborate with collaborators in the other categories (the difference is always significant at 1% level). As we have already discussed, collaboration is linked with more international mobility over the previous three years – but not all types of collaboration. Collaboration with researchers from their own institution is linked to slightly less international mobility and less willingness to leave their country of residence. Researchers who collaborate within their own country are slightly more internationally mobile (36% versus 33%), but mobility is much higher when respondents collaborate in other countries and outside Africa (49% versus 32% and 44% versus 28%). Mobility during studies for highest qualification appears to have an impact, since the percentages of respondents who went abroad during their studies are higher in the group who collaborate more with those in other African countries (52% versus 36%) and with those outside the continent (46% versus 34%).

Gaining more insight into the nature of collaboration

Various issues around research collaboration were addressed during the qualitative interviews with our respondents. These included questions such as: Why do they collaborate? Conversely: Why do some report that they do not collaborate? If they do, who do they collaborate with? And what strategies were pursued in collaborating with others? And finally, what were their experiences with collaboration?

Why do young scientists collaborate?

A scientist from Uganda states that it is necessary to collaborate at the start of a prospective career in research in order to obtain research funding.

> So most time, it's researchers like us who are just starting in the field. You collaborate with somebody, you get a hold of somebody from another university, they write a proposal and you write with them together. So in the beginning, that is mostly how you get your research funding, through somebody, not directly. But going directly would be a big challenge. (35-year-old male from Uganda)

Numerous interviewees indicated that they have to collaborate in order to fulfil the stipulated criteria for research funding.

> Normally when you apply for a grant they expect you to have an already existing collaboration. So, you can't go and look for a collaborator when you do the application: you have to have a track record with collaboration or with an American researcher already in place. So, what actually happens is that in the end you have forced collaboration, that you need to seek collaborators abroad. And it's frowned upon when you do research in-house completely. (Male academic from South Africa)

> My scheme, it basically requires the partnership I have going on with the sandwich programme because I've been looking more into novo diagnostics that hasn't yet been applied on the continent. The core channel for this would be an association with established research institutions and researchers probably from western institutions that are working on new diagnostic approaches. The primary was competitive, but their research associateship also based on an ongoing project. (35-year-old male from Tanzania)

> Most of the research grants now encourage collaborations between departments and universities across the region. So there are a lot of collaborations involving different universities. (35-year-old male from Uganda)

Other interviewees reported that, although not a funding criterion, collaboration may increase the probability of obtaining funding.

> Our vice-chancellor encourage[s] us to do collaborative research, in our faculty and institution, because if you are going pursue research on your own, it can be difficult, and you may struggle to find funding and may even struggle to publish. However, if you work with other researchers, then it gets easier to get funding and publish. (39-year-old female academic from South Africa)

Funders may prefer to fund research of researchers who have a range of skills and experience.

> So when [we] collaborate in terms of bidding for a tender, we can bid with a particular university because we want to show the client that we do have different but also strong skills in this field. Therefore, we are able to carry this research, and also we have done one, two, A, B, C research projects. (35-year-old male from Zimbabwe)

Collaborating with other researchers (and students) helps increase productivity.

> Actually, the publishing one is a result of the collaboration which I already explained to you … As I already indicated I'm working at a university of which you need to write papers … It's a result of collaboration which pushes

> you to write papers and to publish. So, that is the result. (35-year-old male from Ethiopia)

> I know people at other universities that, the moment the opportunity arises, they will contact me and not because I'm a brilliant scholar, but because there is collegiality amongst us and together you achieve much more than what you achieve in your little silos. (Female academic from South Africa)

Opportunity to learn

Partners within a research team have individual strengths and collaboration therefore presents an opportunity for learning.

> I've discovered, when you collaborate, you work as a team and as an individual, you also improve yourself. In that sense, I have discovered that whenever we try to write a paper with some colleagues to exchange the paper, you tend to learn something. Like in my field, we do quite a bit of quantities, analysing things in a quantitative way. So you tend to discover that maybe, a certain colleague is very good in terms of the writing, the technical language. Then somebody is really good in terms of just analysing things using certain software or trying to do certain complex analysis, maybe perform, you compare certain models, someone is very good. So you tend to learn something. Even the way you organise your work, you're now publishing a paper, someone would say, you have to put these things in this way. You tend to learn a lot, and it also improves in terms of time. (35-year-old male from Zimbabwe)

Typically, more senior and international partners are more skilled.

> I'm usually a co-investigator, and so many other people are co-investigators ... So they say the principal investigator has to be a UK citizen, and that already means that it's not balanced. But I do not feel, as a person who goes onto a project, I do not feel that it's unbalanced. I feel that my views are heard, my contribution is appreciated, and I have the right to voice my opinions and contribute. But, you also have to remember that coming from a low- and middle-income country, developing country, the skillset, they are more skilled than I am, so probably their contribution is more than I do. But I also have cultural knowledge, the thing that I bring to the table. So, I feel that at so many levels we are equals on the project. In terms of contribution, it's easy to tell. (34-year-old male from Uganda)

> I collaborate with guys who are more senior than me here in South Africa and abroad ... So those are more senior guys as me, but we work together, we do a project together, we publish together ... I know I'm collaborating with the guys who are senior, so those collaborations I'm learning a lot of things and I can publish a lot. (41-year-old male from South Africa)

To subdivide research activities and share resources

Collaboration allows researchers to focus on preferred research activities.

> We rely on colleagues from other departments ... I work currently now with colleagues in Kenya. Of course they do collect the data there, we do what we do, we work with the proposal, then they go do collect the data, then we write up together, in fact part of my interest mostly is to write up and analyse a report, so I enjoy that. (42-year-old male from South Africa)

Interviewees reported collaboration resulting from the sharing of lab resources.

> I work with different departments in my faculty – nursing and medical technology as well ... I've got a master's student who's dependent on that lab because my department, don't have a lab, and we talked about things like supervision and publications – obviously they are bio-technology and they are going to publish more on the technical side of things and not from an environmental health point. So, it's a win-win situation, once you've sat down around the table and you've discussed what are the options, you reach an agreement on what can be done and how to work together ... and then there's the opportunity for skills transfer as well ... There's a win-win situation for everyone that's involved, but I think the community has to be more open to these kinds of collaborations. (39-year-old female academic from South Africa)

To promote interdisciplinary knowledge

Collaboration across disciplinary lines also promotes the development of interdisciplinary knowledge.

> Problems cannot be solved by single discipline, for example in animal sciences, I'm a geneticist, and the economist should be required, social personnel are required, nutritionists are required. So this has to be collaborative in order to win this large funding to produce maybe valuable research articles. (38-year-old male from Ethiopia)

In some instances, young scientists were approached for collaboration. This was due to a referral made or based on publications.

> They just went onto the Net and Googled; they were looking for someone who has done gender and water in the rural areas. And they called me, and they just hand-picked me and gave me a consultancy. And there are even some people who called me in Tanzania for a conference, also based on my publications. They just went onto Google, got my publications and contact, called me, asked me to write a paper. And that paper was actually published. So, I think all that happened because these people were able to get onto my publications which are on the internet. (38-year-old male from Uganda)

> The collaborative project specifically came my way because a senior researcher I know and who knows about what I did in my PhD knows the researchers, other researchers in the project, and when they contacted him and asked him if he knew about someone who worked on the topic, he referred them to me. (29-year-old female from South Africa)

No collaboration: Reasons and barriers

Inability to find partners for collaboration

Despite various attempts some interviewees are unable to secure partners with whom to collaborate.

> So there is one we've been trying to apply, but currently we haven't been able to get a collaborator. Sometimes, it's also not easy. You write to someone, someone tells you that he's not available or he doesn't have an interest in that area or at the moment, he's a bit busy. It's also not very easy. You can't say that it's very easy. (33-year-old male from Kenya)

This is particularly the case with regard to international collaborations.

> Within the country not so difficult. But outside the country, apart from those I made on ResearchGate and LinkedIn, yes. Those are the few places I have access to. Outside that, like linking up with institutions in my field has been a challenge. (35-year-old female from Nigeria)

> There's very little people to network with in my own field, locally ... And internationally, I haven't really made the contacts internationally, you know, for people who I know I can approach, you know, can you help examine the work, but that it slowly improving internationally, our relationships internationally. It's difficult and a bit daunting to approach the experts in the field and, you know, but it is something that I'm doing, but locally, there's a very, very small research network. (28-year-old female from South Africa)

Lack of resources

An interviewee from South Africa feels that without the required equipment it is very difficult to induce collaboration.

> So I feel the moving off to start moving to collaboration with other profs is definitely one of the key things, and then also profs with other, you know, in other universities ... That whole process is a challenge, yes. And sorry, the other point I just want to make though, is to form collaborations with other professors, I would have to do my experiments on their equipment, because I won't have any of my own equipment. Because it's all my supervisor's equipment. So that's the real sort of technical problem we have. If I want to [start] collaborating or try and collaborate with other people, what can I offer

> these other people, other than my man hours, because the equipment I don't own? (31-year-old male from South Africa)

Funding

Without sufficient funding, young scientists are unable to maintain a relationship with partners and undertake collaborative research.

> The limitations mean that there is no funding. All the funds that the university has given us, I have used my cash, because I have to pay, and I have used my cash … without the university, without any funders who have contributed to my accommodation. So, that kind of relationship is extremely difficult to maintain. Whenever they need me, it is over a thousand dollars for the ticket, and then some of that money is to maintain myself … you can't go anywhere, unless you are willing to sacrifice your salary. (34-year-old male from Uganda)

The lack of funding to attend conferences adversely impacts on emerging scientists' ability to network with potential research partners.

> We do apply to attend conferences even outside the country, but you are told, you can't fund the air tickets … So funding the air tickets, you know that is almost three quarters the cost. So it brings a challenge in that kind of context, it brings a challenge. So you find that for you to get much more information from other colleagues from conferences, for me sometimes it's difficult. (40-year-old male from Kenya)

> Funding. Most of the time it's the funding because the university doesn't afford the international conference, so most of the time, it actually supports the local one. So for the international one, I have to look for the funds out there. So if I'm not lucky, I won't be able to attend. (33-year-old male from Kenya)

Funding is also required to attend meetings.

> I need also to network, I need to find some ways to start collaboration with us. Now being stationary, how is this going to happen? For example, I've been talking about this [upcoming] meeting which will be done in Malaysia, it's very important. But because of funding, there is no funding, how am I going to reach there, I can't. (39-year-old male from Tanzania)

Language

An inability to communicate in the language of potential partners is another factor that inhibits collaboration.

> Again, another problem I have with Ethiopia is their national language is Amharic, and the processes are in Amharic too. Communication is a major barrier. When you are in a situation that you want to tell somebody something

> and you need to talk to them through an interpreter, the context of what you are saying might be lost or the interpretation might not be the right thing that you want to convey. All those things create a number of challenges when language is a barrier. (38-year-old male from Ethiopia)

Institutional barriers

Some institutions may not grant the required leave of absence to undertake collaboration.

> Some organisations, believe me, especially in countries that are still underdeveloped or still developing, they don't even allow scientists that opportunity to go and present their papers. Or to just go and to attend certain training to develop their new ideas or improve their skills. Something like that, they don't allow, and sometimes, they can say, like, we don't have travel allowance for that thing. So that is a challenge, where you can meet your colleagues, peers, you exchange ideas, you exchange notes. Sometimes, they are motivated, you collaborate, that thing of travelling from like Zimbabwe to go to attend a conference in America, in Japan. (35-year-old male from Zimbabwe)

At a number of institutions, criteria for promotion and performance reviews discourage collaboration.

> The university promotion rules do not encourage collaboration because if I publish alone, I score higher marks. If we have two, they divide the marks into two and if you are three they divide it, so you can see that that will not encourage anybody to collaborate with another ... But I do do collaborative work with people and even colleagues from other universities, but a person that, if you do that you are on your own because if you collaborate with anybody and it is published you have challenges of scoring very low in that area, no matter how good that work is ... That that will not encourage anybody to collaborate with another. (40-year-old male from Nigeria)

> We've also had unfortunately misleading management where, you know, on the one hand, one year you'd be told that you should co-author. And those are the articles that would be given, you know, priority status in a sense in terms of performance reviews especially. And then you were told, no but why haven't you got a solo publication for example. So, those are some of the misleading things and then in terms of popular journals and magazines versus accredited journals and magazines, where you know at some point you're kind of almost encouraged to do it you know as if it was career advancing and yet it wasn't, it isn't. (Female academic from South Africa)

With whom do young scientists collaborate?

Young scientists collaborate with various partners. These include:

Fellow students

> I have a colleague … we started doing our PhD together, she was a full-time AERC student and we became very good friends and we have collaborated, I think, on previous such papers. I think one has been, one is published at the *International Journal of African Development* at the University of Michigan. And we've also done one other one that we expect to come out in a book chapter, and then we have another one that has been submitted, it's under review at the *World Development Perspectives* and we have few other things that we were planning on doing this year. (34-year-old male from Nigeria)
>
> I've collaborated with colleagues who I've studied or worked with on one project or another and it's a bit easier for me. Because at least I will know that they will reply even if they don't have time they will still give me some of their time and that's it. (40-year-old male from Uganda)

Colleagues in workplace

> Well, I can attribute that to maybe the place I work. It's in my workplace. So where I work … we have experts in the field of science, technology innovation, entrepreneurship, management. So we'd work in teams. So it makes it a bit easier for us to do it together. (38-year-old male from Nigeria)

Colleagues in other departments

> Multi-disciplinary approaches yes, it would be because I am a … medical scientist and I am able to collaborate with some clinicians and public health scientists in my institution. I am also about to collaborate with another epidemiology study clinician in the neighbour, in a sister university whereby we share equipment. (38-year-old male from Nigeria)

NGOs

> But sometimes you can even collaborate with NGOs. NGOs are apparently beginning to get interested in publishing. Yes. So some of them have projects, and then they invite you and they want you to help them to publish some of their work. So, I think … that is how it happens. (38-year-old male from Uganda)

Industry

> I need industry engagement and the marriage of the two is where I find my research. So, I have extremely close links with industry and it's like, they literally pick up the phone and say, you know this is the story here, can we get research conducted on this. Where do we stand on this one, do you know something about this problem? And [I say] no I don't, let me get some students involved in this and let's see what we can find out. (38-year-old female from South Africa)
>
> But also with industry – we partner with industry, we do a lot about that also and that's our focus, private partners – we call it PPP, Private Public Partnership – so that's what we are encouraging now. (40-year-old male from Tanzania)

Only local partners

> Well, at the moment I don't really have any collaborator [overseas]. I have been contacting people, but I have not been getting any feedback at the moment. (35-year-old male from Nigeria)

Preferred partners

Non-local partners

Collaboration with international partners is preferred by some interviewees due to a perception that there is then less rivalry and greater expertise.

> The model in our field, in management studies is really that you collaborate and that's how you get better projects done but I don't know how always, I don't see that much here. I feel there's more of a tendency to collaborate internationally than locally because locally I still feel like a lot of people are protecting their space and some of the more successful collaborations internationally where I think they're not scared to do that or they don't feel threatened to share their stuff with someone else and work together. (38-year-old female from South Africa)
>
> So it's easier to collaborate with people outside because they are at the top of their field, they know what's going on currently in that area, in that research field than back here in Nigeria … But here in Nigeria, because people don't have access to current books, so they are not as on top of the field. (35-year-old female from Nigeria)

Industry

Links between industry and academia should be strengthened to facilitate collaboration.

> But now in terms of maybe taking some time to be out to really meet the

> industry and really appreciate some of the challenges that are happening to the industry, such opportunities aren't there. The academia and industry linkages are still not strong. I think that it's stronger than in Kenya, but it is still an issue for me. (40-year-old male from Kenya)

Strategies to facilitate collaborate

Set up laboratory/centre

A young scientist in Ghana has decided to establish a laboratory with the goal to become a preferred partner for collaboration in the region.

> I decided to set up a research lab that will bridge that gap … So, what I plan doing is that in the French research labs sometimes they develop products or change how things work and all those kinds of things, and so we can also be in a form, which is backed by collaborating with them, in a form of being a test base for their research labs, so that we can distribute some of those in the Africa region, work with them and they can collect the data and stuff like that. So, that form of collaboration, that form of starting with them, will give us more opportunity in order to be a credible … site. The idea is though not to get people to give the money, but to work on a project that will benefit both countries … For instance just in case somebody is developing or somebody is trying to study maybe climate change in Ghana and at the same time want to know in South Africa and that means we can help in deploying the sensors, collecting the data, as well we can build some of the things here and deploy the sensors, hook it up to the web service and then we work on that, kind of, collaboratively. So, while they are also using it for their African Continent Committee for Global Research, we can also use that data for our own research for monitoring the national climate and stuff like that. So, those are some of the research that I'm interested in. (31-year-old male from Ghana)

Similarly, the establishment of a centre in Nigeria has led to numerous opportunities for collaboration.

> Actually, my department right now, my unit which I am heading, we are working on developing a national microbial culture collection centre … which we don't have in Nigeria as of now … So right now we're working with universities in Nigeria and academicians from universities, and we are even working with private sector industries and all that to actually get all these things together. So we are doing some collaborative work, actually. (39-year-old female from Nigeria)

Become a member of a group/association/online community

An interviewee reports that his affiliation with the African Economic Research Consortium has resulted in various opportunities to network.

> So the thing with the AERC is, you have them bring different people, different resource persons from all over the world, including Africa as well. So these people come and teach different courses and then doing the biannual conferences twice a year. And, you know, once you go for this conference you can invite resource persons, all resource persons you can invite, we invite anyone, you know, you're privileged to meet like Stiglitz, you know, renowned people that's there who you can interact with. Yes, so there AERC is doing it a lot and of mentoring and encouraging ... it gives you that opportunity to meet a lot of people, everyone, you know, you never know who you're going to meet in the AERC biannual you have. (34-year-old male from Nigeria)

Being part of the UN-Habitat Partner Universities has also resulted in opportunities to collaborate with partner institutions.

> My university belongs to the UN-Habitat Partner Universities and we've done some work with other universities on that platform. We also belong to the Association of African Planning Schools and that has kind of expanded our reach, at least on the region, and the African Urban Research Initiative has opened up opportunities for collaborative work. (40-year-old female from Nigeria)

An interviewee from South Africa cites that membership to various organisations as well as being a journal editor has resulted in making various research connections.

> I already have my own research network. First of all, because I'm a member of a couple of organisations, like Afrilex, LSSA, and the African Language Association of Southern Africa ... And yes, also one that also makes me have quite an interesting network just because of the editing ... because I have been on the editorial committee of [a journal] for the past five, six years. And I've been a member of it because we're actually free editors for the past two years now, 2015, I've been a member of the department of free editors. So then, this year I was responsible for editing the journals so I was happy to have contacts, not only in South Africa, but also outside the country. (42-year-old male from South Africa)

A scientist who develops software, reports that being part of an online ccommunity has led to collaboration.

> I contribute to open source software and in that respect I've been able to form a local community and kind of get into the local community ... So I've managed to find my people among those. That's a combination of industry and academics. So that's a useful position. So I think in terms of local networks that is the largest one that I have and that's been going now for three years where I've spoken at a few events and so on. (39-year-old male from South-Africa)

Conference and workshop attendance

Not surprisingly, though, it is conference and workshop attendance that is the most effective strategy to develop research networks and foster opportunities for collaboration.

> Okay, pre my PhD, as a master's student and also as a PhD student there were conferences that I attended on an international basis and that's where I got to interact with other scientists from other countries so from that that's how I built my networks, then I made sure that I keep in contact with them and I also attend workshops or [unclear] that build on my career growth and that's where I also meet some of them and get further information on conferences which are upcoming. (29-year-old female from South Africa)

> I would say that I've established a research network because I know every year I try to attend one conference or the other. I attend conferences in the United States most especially so I have some recent networks over there. And I think I have also, I mean connected some of my colleagues here to some researchers outside Nigeria for their work, for their research work. So I don't really have issues as regards to that because I had the opportunity to, you know, meet scientists out there and mix minds, you know, exchange ideas, I'm in contact with them. Sometimes I've been in contact with them. But so many of my colleagues here don't have that opportunity ... Some of those that have gone out of the country they come to us for advice and connections to other research, to other people outside Nigeria for collaboration. (39-year-old female from Nigeria)

Attendance of international conferences are considered preferable for fostering future collaborations.

> Travel overseas to conferences. Then it's relatively easier to network and get colleagues that are working on similar fields far away, and you will collaborate, so that's the kind of work I will do personally. I don't know how other guys do, I think it's similarly, they have collaborations with other colleagues, within the university and with other universities, that they have relationships with. (42-year-old male from South Africa)

Student network

Student networks are also useful, particularly to foster collaboration with industry.

> So the teaching has this one beneficial effect that I've become known in the industry because my students are everywhere, and I've had very good success with approaching industry for problems to solve and solving their problems. (39-year-old male from South Africa)

Invite scholars

Another strategy that may result in collaboration is to invite scholars to a local institution.

> If we were to invite visiting scholars but it doesn't end there, we maintain a relationship with those scholars because it's not like if you are all employed in one department that you've got the same research interest. (41-year-old female from South Africa)

Negative experiences/ consequences of collaboration

Loss of research autonomy

Numerous interviewees reported negative collaborative research experiences. In some cases, there may be less autonomy to conduct research and the research focus may have to be adjusted.

> I think the main challenge is, for a researcher in a developing country, the problem is we don't have active funding. Most of the time you have to rely on collaborations in the workspace for funding. I must say that is a big limitation, especially for within the broken work setting. South Africa is a bit different, but where I was before in Tanzania, I think it's a little bit tricky because you're the only driver in the field, but not really in that aspect of your research. (40-year-old male from Tanzania)

> Well, I'll just be blunt, most of the time we have to tweak our research concepts or proposals to fit with their agenda ... The focus, the objectives, because we have to work in collaboration and in more cases than the opposite, we find that we have to redesign our objective to fit with the more advanced environment and sometimes we lose focus, we lose our objective. Most times we present the research we've got in our home country, so our own research, the impact or the value of that research is lost because of the way the research is conducted. (35-year-old male from Tanzania)

Junior partners may be powerless to decide what to do with the research funds made available.

> Disadvantages associated with collaboration. If I take North–South cooperation, for example, the money does not come from the South partner. He is in a weak position to negotiate anything. He needs the remainder of the budget that will be given to him to implement his work. It also needs that, and the data, often when it's high-level studies, early writers ... negotiating ranks of writers, given that you who are in Canada you have more access to high-level literature level, you have more access to scientific news. In terms of writing and everything, these people are ahead of those in the South. And if the order of the authors in a scientific publication must be done according to the contribution of each author, you see very well that those of the North are well in advance compared to those of the South. What makes those in the South will occupy, what I call, additional staff. (Male respondent from West Africa)

> Definitely, because there will be some senior, or principal scientist that you will be working under and most of the time, when the funds come, they control the funds. So, maybe there's a project, maybe the senior scientist, or the principal scientist will be the principal investigator and you are supposed to work under him, or her. So, in that case, you don't have true access to some of the exposures. Maybe there might be training. He will decide who to go and those things. So, unless that person is, let me use the word, and then maybe that person is kind and reasonable. Sometimes they keep most of the opportunities away from you, the young scientist. (32-year-old male from Ghana)

Local institutions are the weaker partner

There are reportedly not enough partnerships with local institutions and more effort is required to ensure that collaboration results in upskilling.

> For example, granting agencies can make it a requirement that if you're doing a study in a country which is not your own, you need to include researchers from that country. So I think it's not really fair for researchers to come to Zimbabwe or South Africa or Zambia, conduct their research, publish it, and then without including, so there's no capacity building. So one way to deal with that problem is actually to make it probably part of the conditions or in terms of funding, that's more my suggestion, and also to reach out. (35-year-old male from Zimbabwe)

> My experience has also been that a lot of academics/researchers from outside of South Africa come and do research in South Africa and then they publish. We are based in South Africa, we need to do the research, write up publish the work. I acknowledge that we may not always have the expertise or leading scientists, in specific research areas, but therefore need to have collaborative research for knowledge exchange and capacity building. (39-year-old female from South Africa)

Unequal distribution of workload

Collaboration may result in an unfair distribution of workload.

> But at times it is difficult to work when you are in group. Because at times when you want to come together another one is not free. Sometimes you find that if you are the co-author, you end up doing most of the work on your own. (39-year-old female from Zimbabwe)

This may occur as partners do not complete their share of the work.

> And, you know, the funniest thing? I don't just do their work … I go there for a month, every day, and they don't even bother. Last time we finished the work; they refused to write the work. We broke the work into parts; you write this, I

> write that. At the end of the day I end up writing both parts. And that is why in most of the papers I write you will see I am the first person. That is my part, I have written my part. (32-year-old male from Nigeria)

Some suggestions and ideas about collaboration

Regional initiatives

There is a need for greater opportunities to collaborate with fellow African scientists.

> African, more African collaboration and discussing our challenges. Because I appreciate that our challenges are quite different from those of the scholars in the western world. And I know that we have issues of, you know, making, you know, collaborations and so on and so forth … So we need internal African collaboration. (37-year-old male from Uganda)

Government-driven initiatives

Nationally coordinated research agenda

It is suggested that the state coordinate and support research efforts undertaken that are of national strategic importance.

> Collaborative research with country-level needs in mind. An example is the lack of data on water and sanitation for SA on the Joint Monitoring Programme, we can perhaps work together with other universities to collect the data across all the provinces and respond to [the] research gap, without necessarily competing. I can create a collaborative research group that can work on water and sanitation across all the universities in different provinces, and then do research through collaborative study done at the same time across the provinces and then I wouldn't need to do a country-level study on my own. So, I think collaborative research can work well, as we do a joint funding application, sign a memorandum understanding and agree on roles and responsibilities, benefits, objectives and outcomes. Creating the platform and opportunities for communication is very important. (39-year-old female from South Africa)

Links with industry

It is considered the responsibility of government to create links with industry.

> I think the greatest help would be Government linking us with industry, especially as up-and-coming researchers … So, I think if government can establish that link … in fact, that's what my PhD is all about, into understanding what is the problem in having these university, industry and government linkages … because sometimes you struggle a lot to get industry to be part of your study. And then even if one does and you send them that document, they just totally ignore that. (30-year-old male from South Africa)

Association-driven initiatives

It is expected that associations should make more effort to link scientists together.

> But you find that the conferences are not taking place, there also isn't any maybe platform in the web site of the Association where we can have maybe research group or just discussions about certain topics. So yes, the Association is there, but for me it's not giving us the opportunity to network with other colleagues, but then at the conferences, also yes, one is able to talk to one or two people, but really, it is not easy, especially as somebody who is new in this particular field, you know. (44-year-old female from South Africa)

Institution-driven initiatives

Individuals to share networks

Individuals within institutions can share their networks and networking opportunities.

> [We need] knowledgeable others who can advise you and who can work with you. Pardon me, it's just so difficult to establish a network if you don't have someone who, who can introduce you to people, who can help you connect to certain people. (33-year-old female from South Africa)

Bring international scholars

International scholars can come to present workshops.

> I mean I think it would be great, again, it comes back to bringing international scholars down, so in different sort of, whether it's different provinces that they could bring, I don't know, different types of methodology people down to just, or even some people in-house and have sort of workshops. (38-year-old female from South Africa)

Between universities

There is also a need for developing and newly established institutions to have opportunities to collaborate with other (more) established institutes.

> I believe that perhaps maybe as colleagues, in this particular discipline, maybe we also have to make initiatives … between two universities, let's say two universities that are offering development studies, then we come together as those two departments, we have seminars where we discuss papers, and perhaps, rather I think that it could serve as a good platform for us to network, because I am also thinking if I meet a colleague, maybe from one university at a seminar, and then that particular colleague may know somebody who is interested in that field that I am also interested in. That colleague should then refer me to that one, in that way, we are then able to build a network, and ultimately, a community of researchers. (44-year-old female from South Africa)

Creation of like-minded communities

Young female academics

The creation of special interest communities, for example for young female academics, can also create opportunities for collaboration.

> Creating a community of young women academics. Like a deliberate community, like maybe, what's this, an app or a website ... And then have seminars where they meet and they present amongst themselves and they push themselves ... So, you create a platform for them to run it and, then, it forces them to even compete amongst themselves without feeling like they are competing. So, if you see that your colleague is going overseas to present [inaudible] and you're doing nothing, you think, what's the problem? So, they also nurture their own community for you. (33-year-old female from South Africa)

Organisations in specific fields

Certain fields require specialist organisations to facilitate networking opportunities and foster collaboration.

> Something that may also be a problem in conducting research in Africa and in terms of collaboration is the lack of strong applications. We need to come up with organisations in specific fields, and these organisations should be supported. (40-year-old male from Kenya)

Creation of platforms/forums/databases

It is suggested that forums and platforms be created to share views and to make recommendations.

> So I'm just hoping that for the young scientists or researchers in Africa, if we could have like a group where we could share our ideas ... If we had a group where we can share our views, maybe that would be better along the line we're actually sharing with the funding bodies so that somebody will see. You could write a proposal and send somehow, and you get a way of getting the funds to do more higher-profile studies. (40-year-female from Nigeria)

> There should be more forums or something where we can go and bounce ideas off or have a database of everybody who's doing a PhD at this moment in time somewhere on the NRF website or somewhere. I don't know where. Where we can talk to each other because I'm sure there's other people who are having this similar issue like me and they also wish that they have someone to talk to. I go on LinkedIn most of the time and put up a question and say, okay anyone on LinkedIn that's also an academic having this issue, can you please talk to me because you like feel so alone you know? Who do you ask, where do you get information from? And generally, get information

> from people who've already been through the process or have done it before, then you go and ask them. (44-year-old female from South Africa)

It is also recommended that greater use be made of existing platforms.

> There are platforms now. You know ResearchGate is there, RUFORUM. I don't know whether you have also heard of RUFORUM. It's an African ... regional university network in Africa. If you have information, you share there. Locally, we also do what we call talking to stakeholders, where you can pass your information. So stakeholder engagement, publications and all that. So these are some of the ways. Every newspaper, news brief. I remember you were invited to speak on an issue on a radio station. So these are all platforms available. And, of course, WhatsApp, Facebook and academia.edu, you can do all this. (36-year-old male from Ghana)

A (national) database of emerging and established scholars can be created to facilitate networking opportunities.

> The NRF could do something where they sort of have a database of emerging scholars and you're part of it and, you know, you can sign up for more workshops also but where they are able to bring down really sort of exceptional, leading people to help you on aspects but also where it draws a crowd that's not only from your university so you get to meet people then work, for the networking, in other universities in your vicinity and I don't know, maybe that could open up some possibilities for greater collaboration. (38-year-old female from South Africa)

> I think NRF is at a position to help with collaboration if they keep a database of who is established in a certain research area and who is emerging, they can connect those two. I know people always say it's very easy, you can initiate your own collaborative links but it's not easy for an emerging researcher to know that there is that avenue that you can use. When there is a platform that is provided for established and emerging researchers to identify each other and communicate with one another, that will be excellent. (36-year-old female from South Africa)

Additional funding

Additional funds are reportedly required to maintain support for established collaborations.

> You need the money to be there so that these collaborations can grow into something meaningful. You probably need tools that can be used to monitor until such a point when maybe someone becomes an accomplished researcher, then they can take up another person to mentor. Otherwise, at the moment, most of these collaborations, some of them die rather than work out. (34-year-old male from Uganda)

Additional funds are also required to attend conferences/seminars/workshops, etc.

> One is having or getting research funding for travels, for conferences, for presentations, to attend seminars and the likes ... because at the end of the day, we know that we will not go for a presentation or for attending a conference. It boils down to, you know, getting access to funding. (36-year-old male from Ethiopia)

> Recently it's been very difficult for me to find funding to travel to the conferences which are aligned with my own personal research interests, and so I have been able to access some funding, but it's been a bit difficult, and that has then made it very difficult for me to collaborate with people and to, you know, remain motivated. (35-year-old male from South Africa)

Summary and conclusions

Our survey results show that, across all sectors, respondents reported that they collaborate most often with researchers in their own institution. The exception is researchers based in international organisations, who collaborate less often with researchers outside Africa, which might appear as a surprising result. Researchers in public and private research institutions, non-governmental and international organisations also report the highest frequency of collaboration within their own institution, as well as outside Africa.

When we account for scientific field, the category of researchers with which respondents collaborate the most changes. Across all fields they tend to collaborate more often with researchers in their own country than in their own institution, but still less with researchers in other African countries.

As far as gender is concerned, no large gender differences emerge with regard to respondents' collaboration with different categories of researchers (at their own institutions; at other institutions in their own country; at those in other African countries, or outside of Africa). Some patterns are notable, however. Women are consistently more likely than men to indicate that they never collaborate with other researchers, regardless of the category of collaborator. But this is especially the case for collaboration with researchers at institutions in other African countries, and for researchers at other institutions in the respondent's own country. Male respondents, on the other hand, are consistently (but only slightly) more likely than their female counterparts to collaborate often or very often with all the categories of researchers, but especially with researchers at their own institution.

The interview results showed again the importance of the link between collaboration and funding. Many interviewees indicated that they embark on collaborative efforts in order to access and raise more (international) funding and in this way to propel their research careers. But, it is a bit of a Catch-22 situation as many also indicated that they require funding to pursue collaborative opportunities. Not only is there often insufficient funding for such efforts, but institutional barriers also make such efforts quite difficult.

Given the average age of our interviewees (39-years-old), it is not surprising that they indicated that they sometimes feel that they are at a disadvantage when participating in international collaborations. And some interviewees made explicit reference to the fact that the locus of decision-making in such collaborations does not lie with them. Unsurprisingly, many respondents called for more support, both at the institutional level and also at the level of mentors to assist them in their efforts to collaborate more.

PART FOUR

Conclusions and recommendations

A tale of two halves

By the end of the previous millennium science and higher education in Africa – according to most indicators – were in dire straits. The cumulative effect of the funding policies of the last two decades of the previous millennium, the huge growth in student enrolments in higher education institutions, combined with continuing political instability in many African countries created a state of affairs which we described in earlier works (Mouton (2008) as the 'de-institutionalisation' of science. Scientific institutions in many African countries were fragile and susceptible to the vagaries of political and military events. They were severely under-resourced, and suffered because of a lack of clarity and articulation of science governance issues (demonstrated by constant shifts in ministerial responsibility for science). In particular, African sciences was hugely dependent on international funding for R&D. The cumulative effect of the brain drain of the 1970s and 1980s meant that a whole generation of senior academics and scientists had been lost. This would have a devastating effect on the ability of many universities to build the next generation of scientists. At a practical level, when enrolments in postgraduate student numbers did begin to increase at the turn of the millennium, there were simply not enough supervisors and mentors for these students.

A new narrative emerged around the turn of the century. In Chapters 2 and 3 of this book, we looked more closely at whether this was simply empty rhetoric or whether there was evidence of the 'rising tide of African science'. Our findings from both standard bibliometric analyses, research output and impact, and of funding acknowledgements, revealed that there are indeed signs of a more positive shift. Our bibliometric analyses showed that over the past ten years, Africa has seen an increase in the numbers of publications, more international collaboration and increased mobility of African scientists. But we cautioned that these more positive trends do not necessarily reflect the impact of deliberate interventions and strategies of many African states. In fact, we emphasised that many of these more positive developments are occurring outside (and even despite) the decisions and funding of science and innovation by many African governments. In particular we pointed to the fact that these positive changes are directly linked to the continued increase in investment by international funders and the accumulative effect of increased international collaboration between foreign scientists and African scientists in multi-authored teams in such fields as high-energy physics, infectious diseases and tropical medicine. Based on their analyses of funding acknowledgements, the authors of Chapter 2 concluded that 'all in all, the whole production of the continent is characterised by the presence of non-African funders, with the European Union, the NIH and the Wellcome Trust as some of the most important examples'.

It is against this background that we conducted our study of young scientists in Africa. Our results reflect this dual narrative: Some of the findings confirm that our respondents are reaping the benefits from increased availability of funding, more opportunities for mobility and hence increased international collaboration. At the same time, the continued legacy of weak institutions, long-lasting impact of brain drain and the general lack of established support structures are also reflected in the frustrations and negative experiences of young scientists about the many challenges they face.

Recommendations

Our recommendations pertain to young scientists that are younger than 40. Only one-third are female, but the proportion of female scientists is the highest among this younger cohort. They are nationals of, and tend to live in, countries in Southern, North and West Africa. Most work in the higher education sector, but to a lesser extent than their older counterparts, and in that sector, almost half hold the rank of lecturer. The qualitative data show that power differentials still exist within the higher education sector. From the perspective of the young scientists (especially those in West Africa), individuals in senior academic ranks (e.g. professors) or in senior management positions (e.g. deans) need to be more approachable, less domineering, and more trusting and encouraging of their younger colleagues' research aspirations. This is supported by the survey results. Young scientists, in particular, experience challenges in terms of human capacity building and professional development (e.g. mentoring, mobility and training).

The majority of the young scientists are qualified in the natural, health or social sciences, but slightly more than a third are not in possession of a doctoral degree. This probably explains why they work, on average, slightly fewer hours per week than their older counterparts, but are more likely to spend that time on (their own doctoral) research than on training or supervising postgraduate students. However, the qualitative data also show that young (and therefore relatively inexperienced) African scientists simply cannot prioritise their own research if they are overburdened by excessively large teaching loads, especially at the undergraduate level.

There seems to be a lack of recognition, at institutional level, of the extremely time-consuming nature of teaching large, undergraduate classes. An increase in marking and administrative teaching assistance is therefore strongly recommended. In addition, it needs to be recognised that the current institutional strategy of allocating large teaching loads to junior, newly appointed staff, rather than to more research productive, senior members of staff (related to the abovementioned, rank-related power differentials), is an unsustainable one over the long term. Even for those young academics who are already supervising postgraduate students, especially at the master's level, the potential for such supervision to contribute to their own research is undermined by efficiency considerations, i.e. the sheer numbers of these students researching diverse topics and who are in need of close supervision.

A lack of human resources underlie many of the more specific challenges in the careers of young scientists, and do not only involve addressing high student-to-staff ratios. Another case in point is the young academics occupying positions of responsibility within their departments and faculties. Although they constitute a relatively small group, they are particularly in need of administrative assistance. More effective and efficient university administration systems are needed to release the research potential of this next generation of leaders in the higher education sector.

In general, the ideal that teaching and research functions should supplement each other seems more like a paradox than an ideal. At both individual and institutional level in Africa, we observe tension, contradiction and even conflict between these two core academic functions, which systematically impact more negatively on the research careers of young, relatively inexperienced (and powerless) scientists. A strong emphasis on quantifiable research outputs is often out of touch with the daily realities with which lecturers and senior lecturers are faced. Although we recognise differentiation and specialisation of both institutions and career-tracks are controversial matters, it may be the only way to substantially address this tension.

Young scientists in particular are further challenged in their careers by a lack of research funding, both when their perceptions and reported funding amounts are compared to those of their older counterparts. In addressing this challenge, our results further show that funding for research equipment (e.g. upgrading of laboratory machines) should be prioritised above, for example, funding for library and information resources. One field that is shown by both the quantitative and qualitative data as being resource-stressed is engineering and applied technologies.

While male respondents (with a few exceptions) reported higher numbers of outputs (irrespective of field and age), gender by itself, as well as in interaction with age and field, do not seem to be highly correlated with reported funding amounts. Rather, age of respondent and field were the strongest predictors of differences in amounts reported.

Inexperience constitutes a major barrier to securing funding, both formally (i.e. funders' requirements of certain qualifications, levels of experience and international networks are simply not met by young scientists), and informally (e.g. young scientists are relatively inexperienced in writing quality proposals, especially within limited time-frames). Such inexperience is especially restrictive in private research institutes and countries where the scarcity of government funding increases the competition for funding to levels that young scientists – who are still developing their CVs and building partnerships – find very difficult to meet. Those who do attempt to apply for grants, allocate substantial amounts of their time, which could have been used for research, on this task. Those who are unsuccessful, have no option but to use personal financial resources to undertake research-related activities that would further their careers.

Although it is understandable that funders would want to limit the risks associated with funding inexperienced researchers, more differentiated funding mechanisms that take into account level of experience may level the playing field somewhat. For example, seed funding earmarked for emerging researchers would at least allow them to increase their experience, to enter the funding 'market', and thereby to address their marginalised position in funding regimes. At the same time, inexperience in fundraising, and specifically writing quality proposals, needs to be addressed through training, mentoring and constructive feedback on unsuccessful proposals. Institutions, mentors and funders could play a strong role in this regard. Recognising that where, and how, to apply for research grants are 'tacit' skills young scientists often lack, would be a useful starting point.

In general, and not only in relation to funding, the need for training and mentoring emerges as one that is relatively specific to young scientists, and slightly more so for those in professional fields, such as the health and engineering sciences, than in other fields. Our results further seem to suggest that the transfer of 'softer skills' – those that would allow young scientists to, for instance, make informed career-related decisions about job opportunities and establishing networks – is required much more than transfer of 'harder skills', such as those involving methods or procedures. It should also be recognised, especially by higher education institutions, that many young scientists may be first-generation academics, for whom the expectations and roles associated with their positions are unclear. Brain drain compounds the problem, with lacunae being filled by individuals who lack institutional knowledge and support structures.

Inexperience with regard to publishing in journals tends to generate a particularly severe level of stress, even more so than inexperience in teaching, and especially amongst young scientists who work in institutions that lack an established research culture. We found that young scientists (and especially females) produced, on average, a lower number of articles in the preceding three years than their older counterparts did. At the same time, as our qualitative results show, young scientists are aware of the increasing pressure on academics globally, to publish, in order to advance their careers and their position of power in the academic hierarchy.

Many of the recommendations made here would indirectly contribute to an increase in young scientists' journal article output, although unintended consequences of an ill-considered emphasis only on quantity of output and impact factors of journals need to be kept in mind. These include a decline in quality of output, a disincentive to undertake research that is creative and/or has local societal impact, and the temptation to publish in 'predatory' journals.

Suggestions to alleviate the stress young scientists associated with expectations to publish are provided primarily by the qualitative data. Providing guidance in identifying appropriate (and non-predatory) journals for publication was repeatedly highlighted. The supervisor's role is paramount in this regard, as are provision of training and the implementation of checks and balances by institutions, and input from journal editors. As with research proposals, young scientists whose papers are rejected by journals would greatly benefit from more detailed, constructive feedback from editors and/or reviewers. Local journals could provide such a developmental service, and thereby a valuable platform for young scientists who are still learning how to publish. Institutions can play a role by streamlining their research approval systems and ethics approval processes, as well as by providing more adequate research policies and guidelines.

We mentioned earlier that power relations between young scientists and their more senior colleagues tend to be hierarchical, and that competition for funding is fierce. In such a context, it is therefore unsurprising that young scientists often struggle to find suitable mentors, that many potential mentors do not prioritise that role, and that mentors are perceived as 'negative' instead of encouraging. It emerged from especially the qualitative data that formal mentoring programmes, which do not place the onus on the young scientist to initiate a mentor–mentee relationship, are required. In other cases, insufficient numbers of established researchers, often in the more interdisciplinary and emerging fields, mitigate against effective mentoring of new academic staff.

One way for individual young scientists to overcome the challenges they face in their careers in Africa, but especially to develop professionally and to access funding, is to become more internationally mobile. More than a third have travelled in the recent past, and they are more mobile than the oldest generation of scientists. Young scientists are also more inclined to report the advantages of studying and working abroad than their older counterparts are, which provides another perspective on where and how African higher education systems are not meeting their needs. In some cases doctoral and further training is simply unavailable in a young scientist's chosen field. In other cases, there is a general perception that an overseas degree is of higher quality and carries more prestige. This perception is reinforced by appointment and promotion committees, but is also supported by young scientists' actual experiences abroad of higher levels of expertise and a greater concentration of experience.

Young mobile scientists rate overseas countries as better than their home country in terms of opportunities for collaboration and funding. Training in the 'softer skills', such as writing research funding proposals, which young scientists clearly experience as lacking in their higher education institutions (see above), seems to be more readily available overseas. These observations are supported by our quantitative results that young scientists who are mobile are more likely to secure international funding.

Overseas countries are also rated as more superior in terms of research resources. Lack of research facilities in many African countries impacts negatively on young scientists' research productivity, limits their skills training, and could also render certain research avenues completely unfeasible. Our qualitative results illustrate the frustration this causes for especially those young scientists who have experienced working overseas with state-of-the-art equipment, well-stocked libraries and even 'basics', such as office space, a computer with internet access, a telephone, scanner and printing paper. Not surprising then, is the fact that nearly 80% of all young scientists either often or sometimes consider leaving the country where they work/reside, and three-quarters are of the opinion that a lack of mobility opportunities may have impacted negatively on their careers as academics or scientists.

The young scientists who seem relatively less able to access the benefits that mobility brings, include women scientists, those in public research institutions and higher education institutions (as compared to other sectors), and those working in Southern and North Africa. Lack of mobility seems to be relatively more prevalent among, and is perceived to have the most negative impact on, scientists in the natural sciences, engineering, and applied technologies and agricultural sciences (as compared to those in the social sciences and humanities). In these fields, the lack of access to state-of-the-art equipment and laboratories that we already detailed above is especially debilitating.

Our results support the recommendation that mobility of young scientists should be supported and facilitated. Our qualitative results indicate that young scientists require more information on mobility opportunities and funding for attending international conferences. However, it should also be borne in mind that mobility may have an unintended effect. Many non-mobile young scientists are doubly disadvantaged by the permanent relocation of scientists to countries outside Africa, as the resulting erosion of local expertise creates major challenges for those young scientists who remain behind (which have already been alluded to above). Preventing such permanent brain drain should therefore be a high priority, also because of the 'brain gain' that returning researchers offer their African research institutions and countries.

Concluding comments

Our study has produced a rich and fine-grained picture of the young scientist and academic in Africa. We have produced findings and evidence that are more comprehensive and up to date than previous studies. By combing multiple methods – bibliometrics, a web survey and qualitative interviews – we have been able not only to gauge the magnitude of recent and current trends, but also to obtain more insider-type accounts of the personal experiences and challenges that young and early-career academics and scientists face.

Despite many positive signs that the conditions for productive scientific research in African countries are improving, it is also clear that many structural constraints remain in place that need to be addressed by the key stakeholders in these science systems. These constraints refer specifically to the continued dependence on international science funding, the legacy of the effects of the brain drain of the previous century and the lack of sufficient support and mentoring programmes and structures that the next generation of scientists require. Our study has shown that is essential that international funding agencies, national granting councils as well as universities take cognisance of these findings and design appropriate interventions to address these challenges.

APPENDIX 1

Research design and methodology

Catherine Beaudry, Johann Mouton, Heidi Prozesky, Charl Swart and Rein Treptow

In addition to bibliometric analyses (Chapters 2 and 3), the study applied a mixed method design for primary data-collection consisting of a web-based survey and a series of qualitative individual interviews.

Web survey

A web-based survey was conducted between May 2016 and February 2017. More than 120 000 questionnaires were distributed through two online survey platforms. When the survey was closed, a total of 7 513 completed questionnaires had been received – arguably the largest survey of scientists ever conducted on the African continent.

Identifying and contacting potential respondents

To identify and contact individuals from our target population, we extracted corresponding authors' emails from the Web of Science and Scopus databases for each article published from 2005 to 2015 with an institutional address in Africa. For Zambia, we also used articles in journals not indexed in the Web of Science and Scopus databases. Other sources of emails included the South African Knowledgebase database, the internet, as well as snowball sampling.

Survey administration

Data were collected via a self-administered, structured questionnaire. It was adapted from the questionnaire used for the Global State of Young Scientists precursor study (GLOSYS) (Friesenhahn & Beaudry, 2014) and for GLOSYS in ASEAN (Geffers *et al.*, 2017). The study was therefore based on a questionnaire (See Appendix 2) that was partially tested in a 2013 worldwide survey (Friesenhahn & Beaudry, 2014). Following this field-testing of the questionnaire, corrections were made and questions were added, but the core of the questionnaire remained relatively unchanged. The questionnaire was re-tested in Indonesia, Malaysia, Singapore and Thailand in a further study in 2015 and then considerably reduced in size to limit the time required to fill the questionnaire. We focused on items relevant to the African context and the literature gaps we aimed to address. We translated the questionnaire from English into French for respondents in French-speaking countries. The questionnaire is divided into 10 sections: educational background; employment; working conditions;

research output; funding; challenges; international mobility; collaboration; mentoring; and demographic background.

To ensure that the survey would run smoothly, we conducted a pilot study in Zambia during May 2016, and we launched the survey in the other countries one month later. The questionnaire was distributed in three waves. The survey was administered via CheckBox for English-speaking countries (and hard copies were also distributed to respondents in Zambia). For the French-speaking countries, we used LimeSurvey. There were minor differences between the two platforms, but the results were merged into one dataset without compatibility issues.

Potential respondents were first asked whether they wanted to participate (Wave 1). Undelivered emails and inactive email addresses were then identified. To the individuals who agreed to participate, we sent an e-mail containing the link to the survey. A reminder was sent a week later to the potential respondents who had not completed the questionnaire (Wave 2). During Wave 3, we sent an e-mail with the link to the survey to all individuals with an active address from which no previous response was received. Data collection ended at the end of February 2017.

Table 1: **Launching dates and response rate for each country**

Country	Total number of emails	Valid emails	Wave 1	Wave 3	Number of responses	Response Rate
South Africa	29 541	22 824	24/10/2016	10/11/2016	2 557	12.37%
Nigeria	12 179	11 235	18/09/2016	24/10/2016	971	9.85%
Algeria	11 560	9 584	20/12/2016	13/02/2017	568	5.90%
Egypt	19 095	16 123	12/01/2017	27/01/2017	532	3.64%
Tunisia	13 304	11 284	16/01/2017	13/02/2017	434	3.80%
Kenya	5 406	3 928	30/06/2016	18/07/2016	345	9.06%
Morocco	7 989	6 434	13/12/2016	13/02/2017	343	5.30%
Ethiopia	2 883	2 374	25/07/2016	05/08/2016	252	11.28%
Uganda	2 579	2 174	15/08/2016	29/08/2016	205	10.48%
Ghana	2 312	1 924	08/08/2016	22/08/2016	187	10.75%
Cameroon	1 808	1 402	21/11/2016	13/12/2016	170	12.10%
Tanzania	2 204	1 738	13/07/2016	24/07/2016	142	8.72%
Zambia	1 457	1 077	28/05/2016	22/06/2016	128	15.61%
Zimbabwe	1 008	877	22/08/2016	05/09/2016	125	16.38%
Senegal	1 111	903	19/10/2016	01/11/2016	120	13.30%
Botswana	853	728	28/06/2016	12/07/2016	87	13.28%
Burkina Faso	771	563	21/11/2016	13/12/2016	85	15.10%
Côte d'Ivoire	883	716	21/11/2016	13/12/2016	78	10.90%
Malawi	824	662	24/07/2016	05/08/2016	63	10.77%
Benin	629	469	21/11/2016	13/12/2016	57	12.20%
Congo	362	292	16/01/2017	07/02/2017	33	11.30%
Togo	223	182	16/01/2017	07/02/2017	28	15.40%
Madagascar	465	336	16/01/2017	07/02/2017	27	8.00%

Country	Total number of emails	Valid emails	Wave 1	Wave 3	Number of responses	Response Rate
Central African Republic, Guinea, Seychelles, Chad, Burundi, Comoros, Djibouti	338	257	16/01/2017	07/02/2017	25	9.70%
Democratic Republic of Congo	202	168	16/01/2017	07/02/2017	21	12.50%
Mali	344	262	16/01/2017	07/02/2017	20	7.60%
Niger	334	272	16/01/2017	07/02/2017	19	7.00%
Gabon	258	202	16/01/2017	07/02/2017	18	8.90%

Note: The response rates reported here are an underestimation because many individuals in our initial list had two or more active email addresses.

Data processing and analysis

After data collection, the French and English datasets were merged and we started data cleaning and (re)coding. We cleaned all responses to open questions and 'other' responses by standardising answers and creating categories. New variables were created for statistical descriptions and analysis, for which we used the software STATA and SPSS.

A total of 7 515 individuals completed the questionnaire. However, it emerged during data cleaning that 737 were not African nationals and a further 1 076 did not provide their nationality. Excluding these individuals resulted in a dataset containing 5 700 cases. Because of missing responses on specific questions, the thematic analyses presented in the different chapters in Section 2 of the book, were conducted with varying number of cases (between 4 900 and 5 200).

Qualitative strand

Identifying and contacting potential participants

The qualitative phase of the study commenced in April 2017, with the identification of potential participants. At the end of the questionnaire, survey respondents were asked whether they would be available for an interview and if so, to provide their name and contact details. A total of 3 295 (57.8%) of the survey population agreed to be interviewed. However, the number of in-depth interviews we could conduct was limited, and we were specifically interested in interviewing young African scientists, with a focus on gender and research output.

Based on the final number of completed questionnaires by the close of the survey date (February 2017), we identified those individuals that were eligible to be included in the sample of possible interviewees. We subsequently purposefully selected potential interviewees on the basis of institutional affiliation (prioritising universities), gender, age and field.

Interviews

It is important to emphasise that these interviews were not stand-alone interviews which would be based on a standard interview schedule template. Based on the process described above, we generated individual profiles derived from the potential interviewees' responses to the survey questions. The end result was that each interview schedule would be unique.

The focus in these interviews was threefold: (a) interviewees were asked to elaborate on their specific responses to the survey as well as any qualitative comments they had made; (b) interviewees were asked to explain and indicate the reasons for their responses; and (c) interviews closed by asking interviewees to suggest/propose specific courses of actions that may be to the advantage of emerging scholars.

The interviews were first conducted in South Africa (in October and November 2017), followed by the remainder of interviews in Anglophone African countries (between April and June 2018) and finally in Maghreb and Francophone countries (between May and July 2018). In total, 124 interviews were conducted in South Africa, 73 in the rest of Africa (Anglophone) and 62 in the Maghreb countries and Francophone West Africa. This brought the total number of interviews to 259.

Data processing and analysis

All interviews were audiotaped and subsequently transcribed. Qualitative coding and analysis, using Atlas/ti, were subsequently performed on the data.

APPENDIX 2

The questionnaire

This survey is part of a study which main purpose is to contribute to a better-informed discussion of how to improve current institutional policies in African countries to support research-career development of their researchers. It will investigate the factors influencing research performance and career development.

The study is led by Prof. Johann Mouton (Centre for Research on Evaluation, Science and Technology, Stellenbosch University, South Africa) and Prof. Catherine Beaudry (École Polytechnique de Montréal, Canada). The full list of contributing scientists can be found in the information letter provided with the invitation to this survey. If you have any questions or concerns about the research, please feel free to contact the project manager, Dr Charl Swart (charlswart@sun.ac.za).

You were selected for the study because you presently have (or have had in the past) an academic- or research-oriented career.

Participation to this study is voluntary and there are no known or anticipated risks. The research involves an online survey, which will take approximately 30 minutes to complete. The questions deal with your work environment and career prospects as well as other personal information (e.g. prior education, demographic information). You may decline to answer any question and you may withdraw from the study at any time without negative consequences. All data collected will be treated as confidential and your and your organisation's anonymity will be protected in any reports or publications produced from the survey. Only the participating researchers mentioned above will have access to the data.

This project has been reviewed by, and received ethics clearance through, the Research Ethics Committees of Stellenbosch University and Polytechnique Montreal.

You may withdraw your consent at any time and discontinue participation without penalty. You are not waiving any legal claims, rights or remedies because of your participation in this research study. If you have questions regarding your rights as a research subject, contact Ms Maléne Fouché [mfouche@sun.ac.za; (+27) 0-21 808 4622] at the Division for Research Development.

By clicking to the next page, you agree to participate in this study in accordance with the conditions set out in this document.

Educational background

In this section, we would like to obtain information about your highest qualification.

EDU.1 What is your highest qualification?
[ ] Three-year Bachelors
[ ] Honours or four-year Bachelors
[ ] Masters
[ ] Doctoral or equivalent
[ ] Other (Specify)

EDU.2 In which field did you obtain your highest qualification?
Field (multiple choice allowed): [<check-list>] *list appears according to the discipline selected – Scopus categories*

EDU.3 In what country(ies) did you obtain your doctoral degree (or equivalent)? (Only if EDU 1 = Cat 4)
Country: [<dropdown list>]
Country: [<dropdown list>]

EDU.4 When was your doctoral degree (or equivalent) granted? (Only if EDU 1 = Cat 4)
Year [ ] (yyyy)

EDU.5 What year did you start working on your doctoral degree (or equivalent)? (Only if EDU 1 = Cat 4)
Year [ ] (yyyy)

Employment

This section seeks to gain insight into your employment status and how your employment relates to your highest qualification. General questions about whether you are holding down multiple jobs and if you are being sufficiently remunerated are also asked.

EMP.1 What is your current employment status? If you hold more than one job, please answer for your main job.
[ ] Full-time permanent /tenured [Explanation: A full-time employee has ongoing employment and works, on average, around 38 hours each week. Permanent employees are employed on an ongoing basis until the employer or employee ends the employment relationship]
[ ] Full-time contract (non-permanent) [Explanation: A full-time employee has ongoing employment and works, on average, around 38 hours each week. A contract appointment means that the employee is employed for a specific period of time or task, for example a 6 or 12 month period, and employment ends on the date specified in the contract]
[ ] Part-time permanent [Explanation: A part-time employee works, on average, less than 38 hours per week, usually works regular hours each week and is entitled to the

same benefits as a full-time employee, but on a pro rata basis. Permanent employees are employed on an ongoing basis until the employer or employee ends the employment relationship]
[ ] Part-time contract (non-permanent) [Explanation: A part-time employee works, on average, less than 38 hours per week, usually works regular hours each week and is entitled to the same benefits as a full-time employee, but on a pro rata basis. A contract appointment means that the employee is employed for a specific period of time or task, for example a 6 or 12 month period, and employment ends on the date specified in the contract]
[...] Casual [Explanation: A casual employee has no guaranteed hours of work, usually works irregular hours, doesn't get paid sick or annual leave and can end employment without notice]
[ ] Self-employed
[ ] Unemployed or inactive

EMP.2 Is your current, main job a post-doctoral appointment? (Only if EDU 1 = Cat 4)
[ ] Yes
[ ] No

EMP.3 Please specify the sector of employment of your main job:
[ ] Higher / tertiary education [Explanation: university (public or private), college of technology and other institution providing tertiary education, or other institution directly under control of higher education institution]
[ ] Research institution (public / private)
[ ] Business enterprise
[ ] Private non-governmental / non-profit organisation
[ ] Other Please specify: [...]

EMP.4To what extent is your main job related to your PhD or doctoral degree? (Only if EDU 1 = Cat 4)
[ ] It is not at all related
[ ] It is only slightly related
[ ] It is fairly related
[ ] It is highly related

EMP.5 Is the annual gross income from your main job sufficient to cover your living expenses?
[ ] No, it is not sufficient
[ ] Yes, but it is hardly sufficient
[ ] Yes, it is sufficient
[ ] Yes, it is more than sufficient

EMP.6 Do you have additional sources of income?
[ ] Yes → 9
[ ] No → PRO.1

EMP.7 Please specify your additional sources of income. (Please mark all that apply. We are aware that this is a personal matter, but it is very important for understanding the career choices of African scientists; be assured that your answer will be treated with utmost confidentiality)
[ ] Other teaching activity
[ ] Consultancy
[ ] Scholarship
[ ] Support from partner / spouse
[ ] Support from parents / family
[ ] Personal savings
[ ] Rental income from property
[ ] Other Please specify: [< open form>]

Research Output

In this section, we aim to establish what research output you have produced, as well as when you published your first paper.

RO.1 Please indicate how many of the following forms of research output you have produced over the last three years:
[ ] Articles (including co-authored) in international refereed or peer reviewed academic journals
[ ] Articles (including co-authored) in national or local (your own country) peer reviewed academic journals
[ ] Books (i.e. monographs and edited volumes)
[ ] Book chapters (including co-authored)
[ ] Conference papers published in proceedings
[ ] Written input to official policy documents
[ ] Technical manuals
[ ] Articles in popular journals/magazines, essays, newspaper articles or other public outreach media
[ ] Reports on contract/consultation research
[ ] Presentations at NATIONAL or local conferences to predominantly academic audiences
[ ] Presentations at INTERNATIONAL conferences to predominantly academic audiences
[ ] Patents (applied for and/or granted)
[ ] Computer programmes (including co-writing)
[ ] Creative / artistic works of art performed or exhibited (e.g. films)
[ ] Others Please specify: [< open form>]

RO.2 When did you publish your first research article in a national/ local refereed or peer-reviewed journal?
Year [ ] (yyyy)

RO.3 When did you publish your first research article in an international refereed or peer-reviewed journal?
Year [ ] (yyyy)

Funding

We would to establish the sources, of any, of your research funding over the past three years with the following questions below.

FUN.1 Have you received any research funding over the past three years?
[ ] Yes
[ ] No

FUN.2 What proportion of this funding was from national and international sources? (Only if FUN 1=Y)
[ ] % National
[ ] % International

FUN.3 Please specify which NATIONAL agencies have funded your research over the past three years: (Only if FUN 1 =Y AND FUN2=Cat1):
[Examples: government, national research foundations, science councils, research agencies or commissions; business firms or industry; private non-for- profit foundations/agencies/charities]
[Specify] [< open form>]
[Specify] [< open form>]
[Specify] [< open form>]
[Specify] [< open form>]

FUN.4 Please specify which INTERNATIONAL agencies have funded your research over the past three years: (Only if FUN 1 =Y AND FUN2=Cat2):
[Examples: European Union, USA National Institutes of Health, Germany's DFG, European Commission, DAAD, Bosch Stiftung, World Health Organization; ministries, science academies other government agencies; business firms or industry; private non-for- profit foundations/agencies/charities]
[Specify] [< open form>]
[Specify] [< open form>]
[Specify] [< open form>]
[Specify] [< open form>]

Working Conditions

Working conditions play an important role in career motivation and development. In the following section, we would therefore like to find out more about how you feel about your workload.

WOR.1 How would you describe your current workload for each of the following tasks?

	Very Light	Light	Average	Heavy	Very Heavy	Not applicable
Undergraduate teaching	[ ]	[ ]	[ ]	[ ]	[ ]	[ ]
Research						
Training/supervising postgraduate students	[ ]	[ ]	[ ]	[ ]	[ ]	[ ]
Administration	[ ]	[ ]	[ ]	[ ]	[ ]	[ ]
Service (e.g. counselling of students or patients, voluntary services within organisation)	[ ]	[ ]	[ ]	[ ]	[ ]	[ ]
Remunerated consultation and/or research	[ ]	[ ]	[ ]	[ ]	[ ]	[ ]
Fundraising	[ ]	[ ]	[ ]	[ ]	[ ]	[ ]
Other, please specify	[ ]	[ ]	[ ]	[ ]	[ ]	[ ]

International Mobility

In this section we are seeking greater clarity on the mobility and mobility prospects of African researchers, scientists and scholars.

MOB.1 Are you a citizen and/or resident of an African country?
[ ] Yes
[ ] No

MOB.2 In which country do you currently work / reside?
[<dropdown list>]

MOB.3 [Only if EDU 1 = Cat 4] Did you complete all aspects of your doctoral (or equivalent) education in what you would consider to be your home country?
[ ] Yes
[ ] No

MOB.4 (Only if EDU 1 = Cat 4) From which of the following sources did you receive funding or other monetary support to facilitate your doctoral (or equivalent) studies outside your home country? (Please mark all that apply.)
[ ] Hosting country (government or national agency)
[ ] Hosting institution
[ ] International organisation

[ ] Private sector
[ ] Home country/institution
[ ] Other, please specify [< open form >]
[ ] Not applicable

MOB.5 (Only if EDU 1 = Cat 4) Do you think you would you have been able to complete a doctoral degree (or equivalent) without studying outside your home country?
[ ] Yes
[ ] Unsure
[ ] No

MOB.6 During the past five years, have you lived or worked in a country other than what you would consider your home country?
[ ] Yes
[ ] No

MOB.7 Do you intend to leave the country where you currently work or study within the next 12 months?
[ ] Yes, permanently
[ ] Yes, temporarily
[ ] No

MOB.8 How much do you agree with the following statement?
International mobility is compulsory if I want to have a successful academic or research-oriented career
[ ] Strongly disagree
[ ] Disagree
[ ] Neither agree nor disagree
[ ] Agree
[ ] Strongly agree

Collaboration

With the following question, we would like to ascertain to what extent you have been working with a variety of other researchers.

COL.1 How often during your career so far have you collaborated in joint research projects with the following groups of researchers?

	Never	Sometimes	Very often
Researchers from your own institution			
Researchers from other institutions in your country			
Researchers from other African countries			
Researchers from countries outside Africa			
Researchers from other disciplines/research fields			
Researchers from the opposite sex			
Researchers from business/firms/private companies			
Researchers from government or government-based organisations			
Researchers from NGOs and organisations representing civil society			

Support and Mentoring

We would like to ascertain if you received support and mentoring during your career with the following question.
Have your received any form of mentoring or structured support during your career so far? [Only if EDU 4 = 2011 or later, and RO 2/3= 2011]

	Never	Rarely	Sometimes	Most of the time	Always
Advice on career decisions	[ ]	[ ]	[ ]	[ ]	[ ]
Introduction to important networks	[ ]	[ ]	[ ]	[ ]	[ ]
Attaining a position / job via direct intervention of a mentor	[ ]	[ ]	[ ]	[ ]	[ ]
Accessing research funding via direct intervention of a mentor	[ ]	[ ]	[ ]	[ ]	[ ]
Family / care-related support	[ ]	[ ]	[ ]	[ ]	[ ]
Training in methodology	[ ]	[ ]	[ ]	[ ]	[ ]
Training in fundraising	[ ]	[ ]	[ ]	[ ]	[ ]
Training in (scientific) writing	[ ]	[ ]	[ ]	[ ]	[ ]
Training in presenting results	[ ]	[ ]	[ ]	[ ]	[ ]

Career development and prospects

[Only if EDU 4 = 2011 or later, and RO 2/3 = 2011]

The following section would give us insight into your career development and career prospects.
CAR.1 How would you describe your career prospects?
[ ] Very poor
[ ] Poor
[ ] Satisfactory
[ ] Good
[ ] Very Good

CAR.2 Have you seriously considered changing to a different position within the same institution, country, or to another employment sector?
[ ] Yes
[ ] No

CAR3 To what kind of position(s) did you consider to change to? (Please mark all that apply.) (Only if CAR 2 = Y)
[ ] Management position in higher education / research organisation
[ ] Academic/ research position within the same country
[ ] Academic/ research position in other country
[ ] Research position in private sector
[ ] Non-research position in private sector
[ ] Non-research position in government sector
[ ] Self-employment
[ ] Other Please specify: [< open form >]

CAR.4 Have you ever taken a career break for more than three months?
[ ] Yes
[ ] No

CAR.5 What was / were the reason(s) for the career break? (Please mark all that apply.) (Only if CAR 5=Y)
[ ] Parental leave
[ ] Care for other dependents (e.g. parents)
[ ] Unemployment
[ ] Health reasons
[ ] Other

CAR.6 What was the duration of the career break?
[ ] Months (Only if CAR 5=Y)

Demographic background

In this section we would like to find more about you and your home life.
DEM.1 Are you:
[ ] Male
[ ] Female

DEM.2 What is your year of birth?
YEAR [ ] (yyyy)

DEM.3 What is your current marital status?
[ ] Married or living in a marriage-like relationship
[ ] Separated or divorced
[ ] Widowed
[ ] Single (and not separated, divorced or widowed)

DEM.4 How many children or other dependents do you have?
Please enter a number in the relevant boxes.
[ ] Number of children/dependents aged 0 to 5
[ ] Number of children/dependents aged 6 to 18
[ ] Number of adult dependents aged 19 or older (e.g. elderly)

DEM.5 Are you involved in taking care of your children and/or other dependents?
[ ] Yes, I am involved in taking care of my children
[ ] Yes, I am involved in taking care of my parents and/or other dependents
[ ] Yes, both
[ ] No

DEM.6 Do you care for your parents or other dependents that are not your children for more than ten hours per week?
[ ] No
[ ] Yes

DEM.7 How is the care-work for all dependents distributed in your family or relationship?
[ ]% me [ ]% partner [ ]% others (e.g. extended family, paid service)

DEM.8 How is general housework distributed in your family or relationship?
[ ]% me [ ]% partner [ ]% others (e.g. paid service)

Follow-up

Would you be willing to be contacted by our team in case we need more clarity on some of your answers? If so, please provide your email address:
E-mail: ________________________@________________________

After analysing the data from the survey, we would like to conduct a few follow-up interviews. These interviews will be selected based on individual responses to the questionnaire, conducted in English via Skype, and will take about one hour. Would you be available for such an interview?
[ ] No
[ ] Yes

If you would be available for an interview, please provide your email address:
E-mail: ________________________@________________________

References

Adams, C.W. (1946). The age at which scientists do their best work. *Isis*, 36(3–4): 166–169.

Adams, J., King, C. & Hook, D. (2010). *Global Research Report: Africa*. Leeds: Thomson Reuters.

Adams, J.D., Black, G.C., Clemmons, J.R. & Stephan, P.E. (2005). Scientific teams and institutional collaborations: Evidence from US universities, 1981–1999. *Research Policy*, 34: 259–285.

Akinsanya, O.O. (2012). The role of women in academics: issues, challenges and prospects. *Journal of Research in National Development*, 10: 136–41.

Aksnes, D.W., Rorstad, K., Piro, F.N. & Sivertsen, G. (2013). Are mobile researchers more productive and cited than non-mobile researchers? A large-scale study of Norwegian scientists. *Research Evaluation*, 22(4): 215–223.

Allie, N., Batidzirai, T., Bell, L., Blankley, W., Clayford, M., Labadarios, D., Leiberum, V., Le Roux, N., Mabovu, B., Makelane, H., Matshaya, Z., Molotja, N., Moses, C., Mustapha, N., Nkobole, N., Ntisa, S.J., Parker, S., Phiri, M., Ritacco, G., Rumbelow, J., Saunders, N., Sithole, M. & Vlotman, N. (2014). *South African National Survey of Research and Experimental Development: Main analysis report, 2011/12*. Commissioned by the Department of Science and Technology, March.

Allison, P.D. & Steward, J.A. (1974). Productivity differences among scientists: evidence for accumulative advantage. *American Sociological Review*, 39(4): 596–606.

Altbach, P.G. (1996). *The International Academic Profession: Portraits of fourteen countries*. San Francisco: Jossey-Bass.

Álvarez-Bornstein, B., Morillo, F. & Bordons, M. (2017). Funding acknowledgments in the Web of Science: Completeness and accuracy of collected data. *Scientometrics, 112*(3): 1793–1812. doi: 10.1007/s11192-017-2453-4.

AOSTI (African Observatory of Science, Technology and Innovation) (2014). *Assessment of Scientific Production in the African Union*, 2005–2010. Sipopo, Equatorial Guinea: African Union. http://aosti.org/index.php/report/finish/5-report/15-assessment-of-scientific-production-in-the-african-union-2005-2010

Arimoto, A. & Ehara, T. (eds) (1996). *International Comparison of the Academic Profession*. Tokyo: Tamagawa University Press.

Arora, A., David, P.A. & Gambardella, A. (1998). Reputation and competence in publicly funded science: Estimating the effects on research group productivity. *Annales d'Economie et de Statistique*, 49/50: 163–198.

Arthur, P. & Arthur, E. (2016). Tertiary institutions and capacity building in Ghana: Challenges and the way forward. *Commonwealth & Comparative Politics*, 54: 387–408.

Arvanitis, R. & Mouton, J. (2018). Observing African research: Analysis and policy brief. European Expert Network on International Cooperation and Development.

AU–NEPAD (African Union–New Partnership for Africa's Development). (2010). *African Innovation Outlook 2010*. Pretoria: AU–NEPAD, Pretoria.

Barrett, L. & Barrett, P. (2008). *The Management of Academic Workloads*. London: Leadership.

Baruffaldi, S.H. & Landoni, P. (2012). Return mobility and scientific productivity of researchers working abroad: The role of home country linkages. *Research Policy*, 41: 1655–1665.

Beall, J. (2012). Predatory publishers are corrupting open access. *Nature*, 489: 179.

Beaudry, C. & Allaoui, S. (2012). Impact of public and private research funding on scientific production: the case of nanotechnology. *Research Policy*, 41(9): 1589–1606.

Beaudry, C. & Larivière, V. (2016). Which gender gap? Factors affecting researchers' scientific impact in science and medicine, *Research Policy*, 45(9): 1790–1817.

Beaudry, C. & Mouton, J. (2017). Young scientists in Africa: Factors influencing research performance and career development (Mid-term report). Stellenbosch: Centre for Research on Evaluation, Science and Technology.

Beaver, D.D. (2001). Feature report. Reflections on scientific collaboration (and its study): Past, present, and future. *Scientometics*, 52(3): 365–377.

Blume-Kohout, M., Kumar, K. & Sood, N. (2009). Federal life sciences funding and university R&D. *NBER Working Papers* No. 15146.
Boardman, P.C. & Ponomariov, B.L. (2007). Reward systems and NSF university research centers: The impact of tenure on university scientists' valuation of applied and commercially relevant research. *The Journal of Higher Education,* 78(1): 51–70.
Borghei, A., Qorbani, M., Rezapour, A., Majdzadeh, R., Nedjat, S., Asayesh, H., Mansourian, M., Noroozi, M. & Jahahgir, F. (2013). Collaboration in research and the influential factors in Golestan University of Medical Sciences research projects (2005–2007): An academic sample from Iran. *Medical Journal of the Islamic Republic of Iran,* 27(3): 101–108.
Boisot, M., Nordberg, M., Yami, S. & Nicquevert, B. (eds) (2011). *Collisions and Collaboration: The organization of learning in the ATLAS Experiment at the LHC.* Oxford: Oxford University Press.
Boshoff, N. 2009. Neo-colonialism and research collaboration in Central Africa. *Scientometrics,* 81(2): 413–434.
Boyer, E.L., Altbach, P.G. & Whitelaw, M.J. (1994). *The Aacademic Profession: An international perspective.* Princeton, NJ: Carnegie Foundation for the Advancement of Teaching.
Bozeman, B., Dietz, J.S. & Gaughan, M. (2001). Scientific and technical human capital: An alternative model for research evaluation. *International Journal of Technology Management,* 22(7): 716–740.
Bozeman, B. & Corley, E.A. (2004). Scientists' collaboration strategies: Implications for scientific and technical human capital. doi: 10.1016/j.respol.2004.01.008.
Bozeman, B., Fay, D. & Slade, C.P. (2013). Research collaboration in universities and academic entrepreneurship: The-state-of-the-art. *Journal of Technology Transfer.* doi: 10.1007/s10961-012-9281-8.
Bozeman, B. & Gaughan, M. (2011). How do men and women differ in research collaborations ? An analysis of the collaborative motives and strategies of academic researchers. *Research Policy,* 40(2011): 1393–1402. doi: 10.1016/j.respol.2011.07.002.
Campion, P. & Shrum, W. (2004). Gender and science in development: Women scientists in Ghana, Kenya, and India. *Science, Technology, & Human Values*, 29: 459–485.
Cañibano, C., Otamendi, J. & Andújar, I. (2008). Measuring and assessing researcher mobility from CV analysis: The case of the Ramón y Cajal Programme in Spain. *Research Evaluation*, 17(1): 17–31.
Chakravarthy, R., Chawla, A. & Metha, G. (1988). Women scientists at work: An international comparative study of six countries. *Scientometrics*, 14: 43–74.
Cole, J.R. & Zuckerman, H. (1984). The productivity puzzle: Persistence and change in patterns of publication of men and women scientists. In P. Maehr & M.W. Steinkamp (eds), *Advances in Motivation and Achievement.* Greenwich, CT: JAI. pp. 217–258.
Cole, J.R. (1979). *Fair Science: Women in the scientific community.* New York: Free Press.
Cole, J.R. & Cole, S. (1973). *Social Stratification in Science.* Chicago: University of Chicago Press.
Cole, S. (1979). Age and scientific performance. *American Journal of Sociology*, 84(4): 958–977.
Cole, J.R. & Zuckerman, H. (1984). The productivity puzzle: Persistence and change in patterns of publication of men and women scientists. *Advances in Motivation and Achievement*, 2: 217–258.
Costas, R. & Van Leeuwen, T.N. (2012). Approaching the 'reward triangle': General analysis of the presence of funding acknowledgments and 'peer interactive communication' in scientific publications. *Journal of the American Society for Information Science and Technology*, 63(8): 1647–1661. doi: 10.1002/asi.22692
CREST (Centre for Research on Evaluation, Science and Technology) (2014). Mapping social sciences research in South Africa. Stellenbosch: CREST.
Cruz, L. & Sanz, L. (2010). Mobility versus job stability: assessing tenure and productivity outcomes. *Research Policy*, 39: 27–38.
Daniel, H.Z., Hempel, D.J. & Srinivasan, N. (2003). Project selection: A process analysis. *Industrial Marketing Management*, 32: 39–54.
Defazio, D., Lockett, A. & Wright, M. (2009). Funding incentives, collaborative dynamics and scientific productivity: Evidence from the EU framework program. *Research Policy,* 38(2): 293–305. doi: 10.1016/j.respol.2008.11.008
Dennis, W. (1956). Age and productivity among scientists. *Science*, 123(3200): 724–725.

Dietrich, A. & Srinivasan, N. (2007). The optimal age to start a revolution. *Journal of Creative Behavior,* 41(1): 339–351.

Duque, R.B. *et al.* (2005). Collaboration Paradox: Scientific Productivity, the Internet, and Problems of Research in Developing Areas. *Social Studies of Science,* 35(5): 755–785.

Ebadi, A. & Schiffauerova, A. (2016). How to boost scientific production? A statistical analysis of research funding and other influencing factors. *Scientometrics, 106*(3): 1093–1116. doi: 10.1007/s11192-015-1825-x.

Farnham, D. (2009). Employment relations in Europe: A comparative and critical review. In J. Enders & E. de Weert (eds), *The Changing Face of Academic Life: Analytical and comparative perspectives.* New York: Palgrave Macmillan. pp. 195–217.

Feist, G.J. (2006). *The Psychology of Science and the Origins of the Scientific Mind.* New Haven: Yale University Press.

Fox, M.F. (2005). Gender, family characteristics, and publication productivity among scientists. *Social Studies of Science,* 35: 131–150.

Franzoni, C., Scellato, G. & Stephan, P. (2012). The mover's advantage: Scientific performance of mobile academics. *NBER Working Paper* 18577.

Friesenhahn, I. & Beaudry, C. (2014). *The Global State of Young Scientists: Project report and recommendations.* Berlin: Akademie Verlag.

Fry, J. & Talja, S. (2007). The intellectual and social organization of academic fields and the shaping of digital resources. *Journal of Information Science.* doi: 10.1177/0165551506068153.

Gaidzanwa, R.B. (2007). Alienation, gender and institutional culture at the University of Zimbabwe. *Feminist Africa,* 8: 60–82.

Gaillard, J. (2003). Tanzania: A case of dependent science. *Science, Technology and Society,* 8(2): 317–343.

Gaillard, J. (2010). Measuring research and development in developing countries: Main characteristics and implications for the Frascati Manual. *Science, Technology & Society,* 15(1): 77–111.

Gaillard, J., Hassan, M., Waast, R. & Schaffer, D. (2002). *Africa (Science report).* Paris: UNESCO.

Gazni, A. & Thelwall, M. (2014). The long-term influence of collaboration on citation patterns. *Research Evaluation,* 23(3): 261–271. doi: 10.1093/reseval/rvu014.

Geffers, J. et al. (2017). *Global State of Young Scientists (GloSYS) in ASEAN: Creativity and innovation of young scientists in ASEAN.* Global Young Academy.

Gieryn, T.F. (1981). The aging of a science and its exploitation of innovation: Lessons from X–ray and radio astronomy. *Scientometrics,* 3(4): 325–334.

Giles, C.L. & Councill, I.G. (2004). Who gets acknowledged: Measuring scientific contributions through automatic acknowledgment indexing. *Proceedings of the National Academy of Sciences, 101*(51): 17599–17604.

Gingras, Y., Larivière, V., Macaluso, B. & Robitaille, J.P. (2008). The effects of aging on researchers' publication and citation patterns. *PLoS ONE,* 3(12): 1–8.

Hassine, O.K.B. (2014). Personal expansion versus traditional gender stereotypes: Tunisian university women. In I. Buskens & A. Webb (eds), *Women and ICT in Africa and the Middle East: Changing selves, changing societies.* London: Zed Books. pp. 81–95.

Hesli, V.L. & Lee, J.M. (2011). Faculty research productivity: Why do some of our colleagues publish more than others? *Political Science and Politics,* 44: 393–408.

Himanen, L., Auranen, O., Puuska, H. & Nieminen, M. (2009). Influence of research funding and science policy on university research performance: A comparison of five countries. *Science and Public Policy,* 36(6): 419–430. doi: 10.3152/030234209x461006.

Horner, K.L., Rushton, J.P. & Vernon, P.A. (1986). Relation between aging and research productivity of academic psychologists. *Psychology and Aging,* 1(4): 319.

Huang, F. (2013). The internationalisation of the academic profession. In F. Huang, M.J. Finkelstein & M. Rostan (eds), *The Internationalization of the Academy.* Dordrecht: Springer. pp. 1–21.

Huyer, S. (2015). Is the gender gap narrowing in science and engineering? In *UNESCO Science Report: Towards 2030.* Paris: UNESCO Publishing. pp. 84–104.

Jacob, B. & Lefgren, L. (2007). The impact of research grant funding on scientific productivity. *NBER Working Papers* 13519.

Jaksztat, S., Schindler, N. & Briedis, K. (2010). *Wissenschaftliche Karrieren. Beschäftigungsbedingungen, berufliche Orientierungen und Kompetenzen des wissenschaftlichen Nachwuchses.* http://www.his.de/pdf/pub_fh/fh-201014.pdf.

Jean-Louis, D. *et al.* (2003). Creating a new articulation between research and practice through policy? The views and experiences of researchers and practitioners. *Journal of Health Services Research & Policy*, 8(2): 44–50. http://journals.sagepub.com/doi/pdf/10.1258/135581903322405162.

Kahn, M. (2018). Co-authorship as a proxy for collaboration: A cautionary tale. *Science and Public Policy*, 45(1): 117–123.

Kasozi, A.B.K. (2017). The status of research at three Ugandan universities. In T. Halvorsen & J. Nossum (eds), *North–South Knowledge Networks: Towards equitable collaboration between academics, donors and universities.* Cape Town: African Minds, pp. 85–108.

Katz, J.S. & Martin, B.R. (1997). What is research collaboration? *Research Policy*, 26(1): 1–18. doi: 10.1016/S0048-7333(96)00917-1.

Kimenyi, M.S. (2011). Contribution of higher education to economic development: A survey of international evidence. *Journal of African Economies*, 20 (AERC Supplement): iii14–iii49. doi: 10.1093/jae/ejr018.

Knight, J. (2008). *Higher Education in Turmoil: The changing world of internationalization.* Rotterdam: Sense.

Kuhn, T.S. (1962). *The Structure of Scientific Revolutions.* Chicago: University of Chicago Press.

Kyvik, S. (1990). Motherhood and scientific productivity. *Social Studies of Science*, 20(1): 149–160.

Kyvik, S. & Olsen, T.B. (2008). Does the aging of tenured academic staff affect the research performance of universities? *Scientometrics*, 76(3): 439–455.

Larivière, V., Ni, C., Gingras, Y., Cronin, B. & Sugimoto, C.R. (2013). Bibliometrics: Global gender disparities in science. *Nature*, 504: 211–213.

Lee, S. & Bozeman, B. (2005). The impact of research collaboration on scientific productivity. *Social Studies of Science*, 35(5): 673–702. doi: 10.1177/0306312705052359.

Lehman, H.C. (1953). *Age and Achievement.* Princeton: Princeton University Press.

Leisyte, L. & Dee, J.R. (2012). Understanding academic work in a changing institutional environment. In J.C. Smart & M.B. Paulsen (eds), *Higher Education: Handbook of theory and research.* Dordrecht: Springer. pp. 123–206.

Leisyte, L. & Westerheijden, D.F. (2014). Research evaluation and its implications for academic research in the United Kingdom and the Netherlands. Dortmund: Technische Universität, Zentrum für Hochschulbildung. Discussion papers 01–2014.

Levin, S.G. & Stephan, P.E. (1991). Research productivity over the life cycle: Evidence for academic scientists. *The American Economic Review*, 81(1): 114–132.

Levitt, J.M. (2011). Are funded articles more highly cited than unfunded articles? A preliminary investigation. In E. Noyons, P. Ngulube, & J. Leta (eds), *Proceedings of ISSI 2011: The 13th Conference of the International Society for Scientometrics and Informetrics.* Durban: ISSI. pp. 1013–1015.

Li, J., Xie, Y., Wu, D. & Chen, Y. (2017). Underestimating or overestimating the distribution inequality of research funding? The influence of funding sources and subdivision. *Scientometrics,* 112(1): 55–74. doi: 10.1007/s11192-017-2402-2.

Maassen, P.A.M. and Van Vught, F.A. (1996*). Inside Academia: New challenges for the academic profession.* Utrecht: De Tijdstrom.

Mama, A. (2003). Restore, reform but do not transform: The gender politics of higher education in Africa. *Journal of Higher Education in Africa*, 1: 101–125.

McInnis, C. (2010). Traditions of academic professionalism and shifting academic identities. In G. Gordon & C. Whitchurch (eds), *Academic and Professional Identities in Higher Education: The challenges of a diversifying workforce.* pp. 147–165.

Melin, G. (2000). Pragmatism and self-organization: Research collaboration on the individual level. *Research Policy*, 29: 31–40.

Merton, R.K. (1968). *Social Theory and Social Structure.* New York: The Free Press.

Merton, R.K. (1973). *The Sociology of Science: Theoretical and empirical investigations.* Chicago: Chicago University Press.

Mohamedbhai, G. (2011). Higher education in Africa: The challenges ahead. *Higher Education Forum*, 8, pp. 23–26.
Mohamedbhai, G. (2014). Massification in higher education institutions in Africa: Causes, consequences and responses. *International Journal of Higher Education in Africa*, 1(1): 60–83.
Mouton, J. (2008). Africa's science decline: The challenge of building scientific institutions. *Harvard International Review*, 30(3).
Mouton, J. & Boshoff, N. (2010) Bibliometric analysis of scientific output. In *African Innovation Outlook 2010*. Pretoria: New Partnership for Africa's Development (NEPAD). http://www.nepad.org/resource/african-innovation-outlook-2010
Mouton, J. & Valentine, A. (2017). The extent of South African authored articles in predatory journals. *South African Journal of Science*, 113(7–8).
Musselin, C. (2002). Diversity around the profile of the 'good' candidate within French and German universities. *Tertiary Education Management*, 8: 243–258.
Musselin, C. (2010). *The Market for Academics*. New York: Routledge.
Nakhaie, M.R. (2002). Gender differences in publication among university professors in Canada. *Canadian Review of Sociology*, 39: 151–179.
Narin, F., Stevens, K. & Whitlow, E.S. (1991). Scientific co-operation in Europe and the citation of multi-nationally authored papers. *Scientometrics*, 21(3): 313–323.
NEPAD Planning and Coordinating Agency (NPCA) (2014). *African Innovation Outlook II*. Pretoria: NPCA.
Nunn, A. (2005). The 'brain drain': Academic and skilled migration to the UK and its impacts on Africa. Report to the AUT and NATFHE.
OECD (2016). *OECD Science, Technology and Innovation Outlook 2016*. Paris: OECD Publishing. https://doi.org/10.1787/sti_in_outlook-2016-en.
Olaogun, J.A., Adebayo, A.A. & Oluyemo, C.A. (2015). Gender imbalance in the academia in Nigeria. *European Scientific Journal*, November: 294–306.
Over, R. (1988). Does scholarly impact decline with age? *Scientometrics*, 13(5–6): 215–223.
Paul-Hus, A., Desrochers, N. & Costas, R. (2016). Characterization, description, and considerations for the use of funding acknowledgement data in Web of Science. *Scientometrics*, 108(1): 167–182. doi: 10.1007/s11192-016-1953-y
Pavitt, K. (2000). Why European Union funding of academic research should be increased: A radical proposal. *Science and Public Policy*, 27(6): 455–60.
Pavitt, K. (2001). Public policies to support basic research: What can the rest of the world learn from US theory and practice? (And what they should not learn). *Industrial and corporate change*, 10(3): 761–79.
Payne, A. & Siow, A. (2003). Does federal research funding increase university research output? *Advances in Economic Analysis & Policy*, 3: 1–22.
Pelz, D.C. & Andrews, F.M. (1966). *Scientists in Organizations: Productive climates for research and development*. New York: Wiley.
Pezzoni, M., Sterzi, V. & Lissoni, F. (2012). Career progress in centralized academic systems: Social capital and institutions in France and Italy. *Research Policy*, 41(4): 704–719.
Pollak, K.I. & Niemann, Y.F. (1998). Black and white tokens in academia: A difference of chronic versus acute distinctiveness. *Journal of Applied Social Psychology*, 28(11): 954–972.
Ponds, R. (2009). The limits to internationalization of scientific research collaboration. *Journal of Technology Transfer*, 34(1): 76–94. doi: 10.1007/s10961-008-9083-1.
Pouris, A. & Ho, Y.S. (2014). Research emphasis and collaboration in Africa. *Scientometrics*, 98(3): 2169–2184. doi: 10.1007/s11192-013-1156-8.
Prozesky, H. (2006). Gender differences in the journal publication productivity of South African academic authors. *South African Review of Sociology*, 37: 87–112.
Prozesky, H. (2007). Gender differences in the publication productivity of South African scientists. PhD dissertation, Stellenbosch University.
Rigby, J. (2011). Systematic grant and funding body acknowledgement data for publications: New dimensions and new controversies for research policy and evaluation. *Research Evaluation*, 20(5): 365–375. doi: 10.3152/095820211X13164389670392
Roach, M. & Sauermann, H. (2010). A taste for science? PhD scientists' academic orientation and self-selection into research careers in industry. *Research Policy*, 39(3): 422–434.

Ross, S. et al. (2003). Partnership experiences: Involving decision-makers in the research process. *Journal of Health Services Research & Policy*, 8(2): 26–34.
Rostan, M. & Höhle, E.A. (2014). The international mobility of faculty. In F. Huang, M. Finkelstein & M. Rostan (eds). *The Internationalization of the Academy*. Dordrecht: Springer. pp. 79–105.
Salter, A.J. & Martin, B.R. (2001). The economic benefits of publicly funded basic research: A critical review. *Research Policy*, 30: 509–532.
Sawyer, A. (2004). Challenges facing African universities: Selected universities. *African Studies Review*, 47(1): 1–59.
Shapira, P. & Wang, J. (2010). Follow the money. *Nature*, 468(7324): 627–628. doi: 10.1038/468627a.
Simonton, D.K. (1984). Creative productivity and age: A mathematical model based on a two-step cognitive process. *Developmental Review*, 4(1): 77–111.
Simonton, D.K. (1994). *Greatness: Who makes history and why*. New York: Guilford Press.
Simonton, D.K. (1997). A predictive and explanatory model of career trajectories and landmarks. *Psychological Review*, 104(1): 251–267.
Smith, D. & Katz, J.S. (2000) Collaborative approaches to research: A Report to the Higher Education Funding Council for England. Available at: http://users.sussex.ac.uk/~sylvank/pubs/collc.pdf.
Soete, L., Schneegans, S., Erocal, D., Angathevar, B. & Rasiah, R. (2015). A world in search of an effective growth strategy. In *UNESCO Science Report: Towards 2030*. Paris: UNESCO Publishing. pp. 20–55.
Sonnenwald, D.H. (2007). Scientific collaboration. *Annual Review of Information Science and Technology*, 41: 643–681
Sonnert, G. (1999). Women in science and engineering: Advances, challenges and solutions. In C.C. Selby (ed.), *Women in Science and Engineering: Choices for success*. Annals of the New York Academy of Sciences. New York: The New York Academy of Sciences, vol. 869: 34–57
Sonnert, G. & Holton, G. (1996). Career patterns of women and men in the sciences. *American Scientist*, 84(1): 63–71. http://www.jstor.org/stable/29775599.
Stern, S. (1978). Age and achievement in mathematics: A case-study in the sociology of science. *Social Studies of Science*, 8(1): 127–140.
Stuart, T.E. & Ding, W.W. (2006). When do scientists become entrepreneurs? The social structural antecedents of commercial activity in the academic life sciences. *American Journal of Sociology*, 112(1): 97–144.
Subotzky, G. (2003). Addressing equity and excellence in relation to employment in higher education. *Edusource Data News*, 38: 1–13.
Tahamtan, I., Afshar, A. S. & Ahamdzadeh, K. (2016). Factors affecting number of citations: A comprehensive review of the literature. *Scientometrics,* 107(3): 1195–1225. doi: 10.1007/s11192-016-1889-2.
Tahmooresnejad, L., Beaudry, C. & Schiffauerova, A. (2014). The role of public funding in nanotechnology scientific production: Where Canada stands in comparison to the United States. *Scientometrics,* 102(1): 753–787. doi: 10.1007/s11192-014-1432-2.
Tamale, S. & Oloka-Onyango, J. (1997). Bitches at the academy: Gender and academic freedom at the African university. *Africa Development*, 22: 13–37.
Tang, L., Hu, G. & Liu, W. (2016). Funding acknowledgment analysis: Queries and caveats. *Journal of the Association for Information Science and Technology,* 68(3): 790–794. doi: 10.1002/asi.23713.
Teferra, D. & Altbach, P.G. (2004). African higher education: Challenges for the 21st century. *Higher Education*, 47: 21–50.
Teichler, U. (1998). Massification: A challenge of institutions of higher education. *Tertiary Education and Management*, 4(1): 17–27.
Teodorescu, D. (2000). Correlates of faculty publication productivity: A cross-national analysis. *Higher Education*, 39: 201–222.
Tettey, W. (2010). Challenges of developing and retaining the next generation of academics: Deficits in academic staff capacity at African universities. Commissioned by the Partnership for Higher Education in Africa.
Tijssen, R. (2007). Africa's contribution to the worldwide research literature: New analytical perspectives, trends, and performance indicators. *Scientometrics*, 71(2): 303–327.

Tsikata, D. (2007). Gender, institutional cultures and the career trajectories of faculty of the University of Ghana. *Feminist Africa*, 8: 26–41.
UNESCO (2017). Women in science. UNESCO Institute of Statistics Fact Sheet No. 43, March.
Van Balen, B., Van Arensbergen, P., Van der Weijden, I. & Van den Besselaar, P. (2012). Determinants of success in academic careers. *Higher Education Policy*, 25(3): 313–334.
Van Honk, J., Calero-Medina, C. & Costas, R. (2016). Funding acknowledgements in the Web of Science: Inconsistencies in data collection and standardization of funding organizations. In *21st International Conference on Science and Technology Indicators-STI 2016.*
Van Rijnsoever, F.J. & Hessels, L.K. (2011). Factors associated with disciplinary and interdisciplinary research collaboration. *Research Policy,* 40(3): 463–472.
Waast, R. & Krishna, V.V. (2003). Science in Africa: From institutionalisation to scientific free market – what options for development? *Science, Technology and Society*, 8(2): 153–181.
Wagner, C. (2002) Science and foreign policy: The elusive partnership. *Science and Public Policy*, 29(6): 409–417.
Waltman, L., Calero-Medina, C., Kosten, J., Noyons, E. C., Tijssen, R. J., Eck, N. J., ... Wouters, P. (2012). The Leiden ranking 2011/2012: Data collection, indicators, and interpretation. *Journal of the American Society for Information Science and Technology*, 63(12): 2419–2432. doi: 10.1002/asi.22708.
Wang, J. & Shapira, P. (2011). Funding acknowledgement analysis: An enhanced tool to investigate research sponsorship impacts: The case of nanotechnology. *Scientometrics*, 87(3): 563–586. doi: 10.1007/s11192-011-0362-5.
Web of Science (2009). Funding acknowledgements. http://wokinfo.com/products_tools/multidisciplinary/webofscience/fundingsearch/.
Welch, A.R. (1997). The peripatetic professor: The internationalization of the academic profession. *Higher Education*, 34: 323–345.
World Bank (2000). Higher education in developing countries: Peril and promise. Report produced for the Task Force on Higher Education and Society.
World Bank. (2011). *World Development Indicators 2011*. World Development Indicators. Washington DC: World Bank. https://openknowledge.worldbank.org/handle/10986/2315.
World Bank (2013, 29 July). Investing in higher education across West Africa. http://www.worldbank.org/en/news/feature/2013/07/29/investing-in-science-and-technology-across-west-africa
Wray, K.B. (2003). Is science really a young man's game? *Social Studies of Science*, 33(1): 137–149.
Wray, K.B. (2004). An examination of the contributions of young scientists in new fields. *Scientometrics*, 61(1): 117–128.
Xie, Y. & Shauman, K. (1998). Gender differences in research productivity: New evidence about an old puzzle. *American Sociological Review*, 63: 847–70.
Zewotir, T. & Maqutu, D. (2006). Gender differences in attributions of success by academics in Africa: A study of the National University of Lesotho. *Africa Insight*, 36: 35–40.
Zucker, L.G., Darby, M.R., Furner, J., Liu, R.C. & Ma, H. (2007). Minerva unbound: Knowledge stocks, knowledge flows and new knowledge production. *Research Policy*, 36: 850–863.
Zuckerman, H.A. (1977). *Scientific Elite: Nobel laureates in the United States*. New York: Free Press.
Zuckerman, H.A. & Merton, R.K. (1973). Age, aging and age structure in science. In R.K. Merton, *The Sociology of Science*. Chicago: Chicago University Press. pp. 493–560.

www.ingramcontent.com/pod-product-compliance
Ingram Content Group UK Ltd.
Pitfield, Milton Keynes, MK11 3LW, UK
UKHW062010041026
15591UKWH00017B/1261